Springer Tracts in Civil Engineering

Series Editors

Giovanni Solari, Wind Engineering and Structural Dynamics Research Group, University of Genoa, Genova, Italy

Sheng-Hong Chen, School of Water Resources and Hydropower Engineering, Wuhan University, Wuhan, China

Marco di Prisco, Politecnico di Milano, Milano, Italy

Ioannis Vayas, Institute of Steel Structures, National Technical University of Athens, Athens, Greece

Springer Tracts in Civil Engineering (STCE) publishes the latest developments in Civil Engineering - quickly, informally and in top quality. The series scope includes monographs, professional books, graduate textbooks and edited volumes, as well as outstanding PhD theses. Its goal is to cover all the main branches of civil engineering, both theoretical and applied, including:

- Construction and Structural Mechanics
- Building Materials
- Concrete, Steel and Timber Structures
- Geotechnical Engineering
- Earthquake Engineering
- Coastal Engineering; Ocean and Offshore Engineering
- Hydraulics, Hydrology and Water Resources Engineering
- Environmental Engineering and Sustainability
- Structural Health and Monitoring
- Surveying and Geographical Information Systems
- Heating, Ventilation and Air Conditioning (HVAC)
- Transportation and Traffic
- Risk Analysis
- Safety and Security

Indexed by Scopus

To submit a proposal or request further information, please contact:
Pierpaolo Riva at Pierpaolo.Riva@springer.com (Europe and Americas) Wayne Hu at wayne.hu@springer.com (China)

More information about this series at http://www.springer.com/series/15088

Sinan Akkar · Alper Ilki · Caglar Goksu ·
Mustafa Erdik

Editors

Advances in Assessment and Modeling of Earthquake Loss

 Springer

Editors
Sinan Akkar
Department of Earthquake Engineering
Kandilli Observatory and Earthquake
Research Institute
Boğaziçi University
Istanbul, Turkey

Caglar Goksu
Department of Civil Engineering
Istanbul Technical University
Istanbul, Turkey

Alper Ilki
Department of Civil Engineering
Istanbul Technical University
Istanbul, Turkey

Mustafa Erdik
Kandilli Observatory and Earthquake
Research Institute
Boğaziçi University
Istanbul, Turkey

ISSN 2366-259X ISSN 2366-2603 (electronic)
Springer Tracts in Civil Engineering
ISBN 978-3-030-68815-8 ISBN 978-3-030-68813-4 (eBook)
https://doi.org/10.1007/978-3-030-68813-4

This Springer imprint is published by the registered company Springer Nature Switzerland AG
The registered company address is: Gewerbestrasse 11, 6330 Cham, Switzerland

Foreword

The Turkish Natural Catastrophe Insurance Pool (TCIP) was established in 2000, following the 1999 Marmara Earthquake. TCIP is a government legal entity responsible for provisioning, implementation and management of Compulsory Earthquake Insurance in Turkey. With its motto "The earthquake will pass and the life will go on!", the TCIP aims to help citizens resume their lives safely after an earthquake. The Compulsory Earthquake Insurance covers structural damage that may result directly from earthquakes and consequential damages such as those from fire, explosion, landslide and tsunami. This way, it helps people returning back to their normal lives by rapidly compensating the loss their buildings may have suffered unconditional to the damage state. The TCIP aims to provide everyone with the Compulsory Earthquake Insurance by offering them affordable premium rates and to increase the insurance coverage with its distribution network of insurance companies, insurance agents and banks. As such, it is very important for TCIP to reach reliable and realistic modeling of potential earthquake losses before the disasters occur, and to develop robust, rapid and scientifically supported methods of damage assessment right after the earthquakes. Considering the recent developments all around the world on assessment and modeling of earthquake losses, particularly after the recent urban earthquakes such as the L'Aquila 2009 and Emilia 2012 (Italy), Tohoku 2011 (Japan) and the Christchurch 2011 (New Zealand) as well as the efforts going towards developing a national loss model, TCIP has decided to organize an international workshop to gather leading individuals and institutions in the areas of earthquake loss modeling and damage assessment. The main objective of this workshop is to discuss the current state-of-knowledge and practice for seeking ways for betterment of available methods and procedures on assessment and modeling of earthquake loss.

The International Workshop on Advances in Assessment and Modeling of Earthquake Loss, organized by the TCIP, invited and gathered the insurance industry representatives, international reinsurance and modeling companies, government agencies and researchers from academia on November 4–5, 2019 in Istanbul. The workshop included country-specific practices in loss assessment following an earthquake and the latest developments in earthquake risk modeling. The workshop is attended by numerous experts from all over the world, including those from Germany, Japan, Italy, New Zealand, Switzerland, United Kingdom, United States as well as Turkey.

The loss modeling and assessment methods developed for TCIP to forecast possible damages after a potential Istanbul earthquake and conduct rapid damage assessment right after the earthquakes are also presented and discussed during the workshop.

This book covers the papers of distinguished contributors to the workshop. I believe this compilation which covers a wide spectrum of modelling and assessment approaches of earthquake loss will form a reliable reference book in the field of earthquake engineering and disaster management, particularly for the insurance sector all around the world.

I gratefully acknowledge the support of organizing committee members of the workshop and editors of this book; S. Akkar, A. Ilki, C. Goksu and M. Erdik. I am also thankful to Mr. I. Gungor, S. Ozturk, M. Basgu and M. Bahar for their tremendous efforts before and during the workshop.

October 2020   Mete Guler
Chairman of Board of Directors
Turkish Catastrophe Insurance Pool
Istanbul, Turkey

Preface

Earthquake disasters present a wide range of physical and socio-economic impacts, with potentially long-lasting effects. These impacts include cascading effects such as business interruption and revenue loss that may lead to financial challenges. Earthquake insurance, where the risks are essentially transferred to international markets, constitute one of the important financial strategies aimed at funding the rapid response, recovery and post-earthquake reconstruction. Earthquake insurance penetration around the globe varies with the risk awareness in the region and also by its affordability. In this connection a rational, risk-based assessment of the pricing of the earthquake insurance has gained importance in recent years. As a result of these risk-based pricing, however, larger rating differentials can enter the market that may need to be smoothed by the insurance sector.

The main tool in the risk-based insurance pricing is the catastrophe (CAT) risk modelling where earthquake events are simulated relying on vulnerability criteria to compute a rational and fair price for the loss to a client's portfolio. The main output of CAT models is traditionally the probability distribution of estimated seismic loss (i.e. EP curve) for a given portfolio. The EP curve constitutes the essential ingredient of earthquake insurance pricing and reinsurance needs assessment process. Although, over the last decades, earthquake CAT modeling has been dominated by commercial vendors, today several local and international scale earthquake risk models are in existence or under development. It is believed that, these new CAT models promise greater transparency.

One of the important issues and challenges in earthquake insurance is the quantification of losses in a rational, fair and expedient manner. Right after an earthquake, it is very important to evaluate the damage level of the buildings in the affected area to assess the financial losses as well as its re-usability and reparability. This process requires an able CAT management, a qualified team of experts in adequate numbers, supported by a rational loss assessment methodology.

For the betterment of the earthquake insurance penetration rate, obligatory insurance can certainly provide a working solution. A number of earthquake pools exist around the world. New Zealand's Earthquake Commission (EQC) and the Turkish Catastrophe Insurance Pool (TCIP) are the two of the oldest and most successful earthquake pools existing. TCIP was established in 2000, after the 1999 Kocaeli earthquake, as a compulsory homeowners' pool to reduce catastrophe earthquake exposure to the government. The owners of residential properties within municipality boundaries are obliged (through administrative measures) to buy earthquake insurance policy for their homes. The penetration ratio of Compulsory Earthquake (EQ) Insurance Policy for Turkey is 56% as of October 2020. The premium rates range from 0.33% (per mill) to 4.14%, depending on structural type, height as of number of floors, and the locality of the building. The maximum insured limit for a single apartment is 240.000-TL and the deductible is 2% of the insured value. Currently, the pool has not enough financial resources for a big EQ in Marmara region, she cedes a large amount of its risks to international reinsurers, which is about 34 billion-TL for 2020–2021 period. Both the TCIP and the Union of Insurance and Reinsurance Companies of Turkey have worked with academia to develop CAT risk models and risk-based pricing for the obligatory and facultative commercial earthquake insurance. In connection with the equitable and timely assessment and payment of post-earthquake losses, TCIP has also developed a new loss assessment methodology which can also be re-purposed to develop pre-earthquake risk assessment and retrofit strategies.

On November 4–5, 2019 in İstanbul, Turkey, the TCIP has organized the International Workshop on Advances in Assessment and Modeling of Earthquake Loss to share recent advances in the loss assessment and modelling of earthquake risk (https://www.tcip-workshops.com/calistay-hakkinda). The workshop was well attended by CAT modeling companies, international insurance/reinsurance companies, state insurance policy makers and the academia.

This book is the byproduct of the aforementioned TCIP international workshop that provided an opportunity to present state-of-the-art discussions on the issues relevant to the earthquake insurance. The papers presented in the workshop and additionally included ones are grouped under three coherent parts that are explained in the following.

The first part encompasses papers on post-earthquake damage assessment.

S. Pampanin introduces a simplified procedure for post-earthquake safety and loss assessment of buildings. In the next three chapters, K. Kusunoki reviews the damage assessment techniques in Japan; R. E. Gonzales et al. discuss the post-earthquake demolition process in Christchurch, New Zealand and; M. Di Ludovico et al. treat the damage assessment in Italy as well as the recent experiences on reparability and repair costs. The last paper in the first part, by A. Ilki et al., elaborates on the post-earthquake damage assessment methodology of the TCIP.

In the second part, earthquake loss modeling and insurance pricing are covered.

M. Erdik discusses the issues on earthquake risk assessment from insurance perspective. H. Crowley introduces the SERA Project and elaborates on the European earthquake risk models. P. J. Stafford reviews risk-oriented earthquake hazard assessment. The seismic fragility and vulnerability of buildings are discussed by L. Di-Sarno and A. El-Nashai. S. Akkar treats on the earthquake loss assessment models and provides a case study on the content loss modeling conditioned on building damage. Finally, M. Kohrangi et al. review earthquake CAT risk modeling with applications to insurance industry.

In the third part, earthquake insurance for resilience and post-earthquake fire are discussed.

F. Bendimerad discusses the role of earthquake insurance in risk reduction and building resilience with specific examples from TCIP. C. Scawthorn reviews the fire following earthquake and the related potential in Istanbul.

These up-to-date authorative papers by the experts are expected to make this book a guide book for the researchers and practitioners in the related fields. We, as the editors of this book, are thankful to all the contributors for their fruitful cooperation. We are also very much indebted to the TCIP administration, for their support and encouragement in realizing this book and the associated International Workshop on Advances in Assessment and Modeling of Earthquake Loss.

Istanbul, Turkey

Sinan Akkar
Alper Ilki
Caglar Goksu
Mustafa Erdik

Contents

Contributors

S. Akkar Department of Earthquake Engineering, Kandilli Observatory and Earthquake Research Institute, Boğaziçi University, Istanbul, Turkey

P. Bazzurro University School for Advanced Studies IUSS Pavia, Pavia, Italy

Fouad Bendimerad EMI, Quezon City, Philippines

M. Comert RISE Engineering, Maslak, Istanbul, Sariyer, Turkey

H. Crowley European Centre for Training and Research in Earthquake Engineering (EUCENTRE), Pavia, Italy

C. Demir Istanbul Technical University, Maslak, Istanbul, Sariyer, Turkey

G. De Martino Department of Structures for Engineering and Architecture, University of Naples "Federico II", Naples, Italy

M. Di Ludovico Department of Structures for Engineering and Architecture, University of Naples "Federico II", Naples, Italy

L. Di-Sarno Department of Civil Engineering and Industrial Design, School of Engineering, University of Liverpool, Liverpool, UK

M. Dolce Italian Civil Protection Department, Presidency of the Council of Ministers, Via Ulpiano 14, Rome, Italy

D. Dowdell Building Research Association of New Zealand, Porirua, New Zealand

A. S. Elnashai Department of Civil and Environmental Engineering, University of Texas at Houston, Houston, USA

K. J. Elwood University of Auckland, Auckland, New Zealand

M. Erdik Bogazici University, Kandilli Observatory and Earthquake Research Institute, Istanbul, Turkey

R. E. Gonzalez University of Auckland, Auckland, New Zealand

O. F. Halici Department of Civil Engineering, MEF University, Maslak, Istanbul, Sariyer, Turkey

A. Ilki Istanbul Technical University, Maslak, Istanbul, Sariyer, Turkey

M. Kohrangi RED Risk Engineering + Development, Pavia, Italy

S. R. Kotha Univ. Grenoble Alpes, Univ. Savoie Mont Blanc, CNRS, IRD, IFSTTAR, ISTerre, Grenoble, France

K. Kusunoki Earthquake Research Institute, The University of Tokyo, Tokyo, Japan

G. Manfredi Department of Structures for Engineering and Architecture, University of Naples "Federico II", Naples, Italy

S. Pampanin Full Professor, Department of Structural and Geotechnical Engineering, Sapienza University of Rome, Rome, Italy;
Adjunct Professor, Department of Civil and Natural Resources Engineering, University of Canterbury, Christchurch, New Zealand

A. N. Papadopoulos Swiss Seismological Service, ETH Zurich, Switzerland

A. Prota Department of Structures for Engineering and Architecture, University of Naples "Federico II", Naples, Italy

C. Scawthorn Pacific Earthquake Engineering Research Center, University of California at Berkeley, Berkeley, CA, USA

Peter J. Stafford Imperial College London, London, UK

M. T. Stephens University of Auckland, Auckland, New Zealand

C. Toma University of Auckland, Auckland, New Zealand

D. Vamvatsikos School of Civil Engineering, National Technical University of Athens, Athens, Greece

Part I
Post-Earthquake Damage Assessment

Chapter 1
Simplified Analytical/Mechanical Procedure for Post-earthquake Safety Evaluation and Loss Assessment of Buildings

S. Pampanin

Abstract The crucial need to develop and implement simple and cost-effective repair and retrofit strategies and solutions for existing structures has been once again emphasized, if at all needed, by the recent catastrophic earthquake events. The significant socio-economic impacts of the Canterbury earthquakes sequence in 2010–2011 as well as of the "series" of independent events within few years in Italy (L'Aquila 2009; Emilia 2012; Central Italy 2016) have triggered a stepchange in the high-level approach towards the implementation of seismic risk reduction, introducing either a mandatory enforcement or significant financial incentives for a national-wide program to assess (and reduce by remedial intervention) the seismic vulnerability/capacity of the whole (non-dwelling) building stock, including safety and expected repairing costs (direct economic losses). This chapter provides an overview of the motivations, challenges and (possible) solutions for such a complex and delicate task with the intent to stimulate awareness, discussion and synergetic actions within the wider international community. Particular focus will be given to the development and on-going continuos refinement of a simplified analytical-mechanical methodology—referred to as SLaMA (Simple Lateral Mechanism Analysis) method—as part of a proposed integrated methodology for either pre- and post-earthquake safety evaluation and loss assessment of buildings, in order to support the engineering community and stakeholders through the various steps of the decision making process of risk (assessment and) reduction.

S. Pampanin (✉)
Full Professor, Department of Structural and Geotechnical Engineering, Sapienza University of Rome, Via Eudossiana 18, 00184 Rome, Italy
e-mail: stefano.pampanin@uniroma1.it

Adjunct Professor, Department of Civil and Natural Resources Engineering, University of Canterbury, Christchurch, New Zealand

© The Author(s) 2021
S. Akkar et al. (eds.), *Advances in Assessment and Modeling of Earthquake Loss*,
Springer Tracts in Civil Engineering,
https://doi.org/10.1007/978-3-030-68813-4_1

1.1 Introduction

The urgency of a medium-long-term plan for seismic retrofit and risk reduction strategy at a national scale is becoming increasingly evident in most of the seismic-prone countries worldwide.

With no doubt the assessment of the seismic vulnerability of existing buildings and the definition of appropriate solutions—i.e. structurally effective, easy to apply, cost-effective, possibly reversible and respectful of the architectural, heritage and cultural conservation requirements—hide a level of significantly higher complexity than designing new structures.

Even more when dealing with damaged buildings in the aftermath of an earthquake event: some of the most controversial issues in the recent earthquakes have in fact been the evident difficulty and lack of knowledge/guidelines in terms of: (a) evaluation of the residual capacity of a damaged building to sustain subsequent aftershocks; (b) selection and implementation of a set of reliable repairing techniques to bring back the structure "at least" to its conditions before the earthquake; and (c) capacity to predict the cost (or cost-effectiveness) of such a repair intervention, when compared to fully replacement costs and accounting for potential aftershock in the near future.

Moreover, as if the technical complexity was not a sufficient deterrent, the constraint of economic resources for a national scale implementation and the lack of a prioritization plan, based on risk considerations, loss assessment and cost–benefit analyses, are often referred to, or blamed as, primary obstacles to the practical implementation of such a broad and ambitious project. Yet, studies and comparative evaluations of the social-economic effectiveness of a seismic prevention strategy, when opposite to a post-event reaction/repair/reconstruction approach, clearly show its significant long-term and national benefits.

To tackle this delicate issue, it is necessary to improve and standardize the tools and procedures ('protocols') for the 'diagnosis' and 'prognosis' of the seismic vulnerability and of the expected performance of existing buildings, in order to estimate, prior to the event, as well as assess, in the aftermath of the earthquake, both the seismic safety as well as the socio-economic impact/consequences/losses.

Such procedures should be based on state-of-the-art but simplified methodologies—preferably analytical rather than numerical approaches—that could highlight the structural weaknesses of the building system, while ensuring consistency of results and proper level of independently from the operators. Similarly, suitable 'therapeutic pathways', i.e. appropriate repair/retrofit strategies, can be defined by comparing alternative options through a cost–benefit approach.

This chapter provides an overview of recent developments at international level in terms of methodologies and regulatory approaches as well as 'diagnosis' protocols and procedures for either pre- and post-damage of existing buildings, with particular attention to (a) the features of the recently adopted national plan for seismic risk reduction in New Zealand, following the recent and devastating 2010–2011 earthquake sequence in Christchurch, (b) the associated seismic assessment and rating methodology, incorporated in the New Zealand Guidelines on Seismic Assessment of Existing Building (NZSEE 2017) and endorsed by the Mnistry of Business Innovation and Employment and (c) the Seismic Risk Classification guidelines introduced in Italy in 2017 (ITA 2017), following two decades of severe earthquake events, and including both safety and economic-loss considerations.

Building on, and merging, these recent developments, an integrated methodology for the evaluation and assessment of either pre-earthquake as well as post-earthquake safety and losses, based on a simplified analytical-mechanical approach, referrred to as SLaMA (Simple Lateral Mechanism Analysis), is presented.

1.2   Seismic Risk Reduction Policies

1.2.1   *The New Zealand Passive Approach—"Before"*

Before the dramatic Canterbury Earthquake Sequence 2010–2011 in New Zealand, the preparation and implementation of seismic risk reduction policies were delegated to the local territorial authorities (TAs). With the exception of the capital city of Wellington, who had initiated an active approach with a trial assessment on a large sample of buildings, the typical policy adopted by the TAs was based on a 'passive' approach, similar to the general trend at international level in major seismic-prone countries. According to such policy, there was no mandatory requirement to assess the seismic vulnerability and capacity of existing private or public buildings unless the owner aimed for a change of use, increase in volume or structural alteration. The Building ACT (2004) had defined a minimum level of 'seismic safety' for a building corresponding to the exceedance of its 'ultimate capacity' under a 'moderate' Earthquake, the latter being defined as *"an earthquake that would generate shaking at the site of the building that is of the same duration as, but that is one-third as strong as, the earthquake shaking (determined by normal measures of acceleration, velocity and displacement) that would be used to design a new building"*.

Translating these legal and policy-making wordsmithing into engineering practical procedures, within the NZSEE2006 Guidelines on "Assessment and Improvement of the Seismic Performance of Buildings in Earthquakes" (NZSEE 2006), the 'ultimate capacity' was taken as the Ultimate Limit State (or Life Safety, using a common international nomenclature) and the 'moderate earthquake' was taken as the seismic demand (spectral intensity) corresponding to 33% of the design intensity for a new building (i.e. 500 years return period for an ordinary structure with a life span of 50 years). A Safety Index referred to as %NBS (% New Building Standard), or Capacity/Demand ratio—given by the ratio of the Capacity at Ultimate Limite State of the existing structure versus the Demand for a new structure—was introduced in the verification method. A building 'scoring' less than 33% NBS would be identified as an Earthquake Prone Building (EPB) with a seismic Rating (or class) "D", thus considered unacceptable (Fig. 1.1).

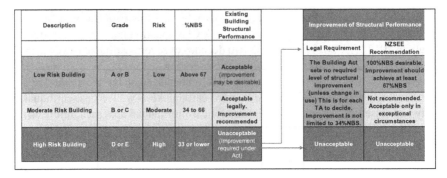

Fig. 1.1 NZSEE2017 guidelines—matrix of seismic rating, vulnerability classes and required (Building ACT 2004) or recommended remedial actions

At this stage, the owner of an Earthquake Prone Building (EPB) was obliged to retrofit/strengthen the building (even just) above the minimum threshold of 33%NBS—at his/her expenses—but only if he/she intended to proceed with the alteration/change of use of the buildings. A potential loophole of such a passive approach– potentially carrying legal liability or at least ethical responsibility—was thus that the owner, either private or public, could opt out and maintain the status quo ("no action" or "do-nothing" approach), in spite of being aware of the high seismic risk of the building and thus of the life-safety risk for the occupants.

It is also important to note here that, as shown in the same figure/table with the same green band, the New Zealand Society for Earthquake Engineering, from its first document NZSEE2006 and furthermore in the significantly revised version NZSEE2017, has been strongly recommending to target 67% NBS as a minimum objective of the retrofit and, as practically as possible, a higher value.

1.2.1.1 Relationship Between %NBS and Collapse Risk

Already the original NZSEE2006 Guidelines were qualitatively discussing the approximate correlation between the %NBS of an existing building and its actual (predicted and approximate) relative seismic risk, intended as a multiplier of the Life Safety (or Collapse) probability of the building when compared to that of an 'equivalent' new building.

Notably, as shown in Fig. 1.2, when decreasing the % NBS—which depends on multiple factors such as: (a) the building material, i.e. unreinforced masonry versus concrete or steel; (b) the structural system, i.e. frame, walls, dual system and (c) the structural details and the design philosophy related to the requirements/recommendations of the design code/guidelines of the time—a more-than-linear increase of the collapse risk, when compared to a newly designed building design, would be expected. Furthermore, the range (dispersion) of the expected Relative Risk would significantly increase, reflecting the higher uncertainties related to

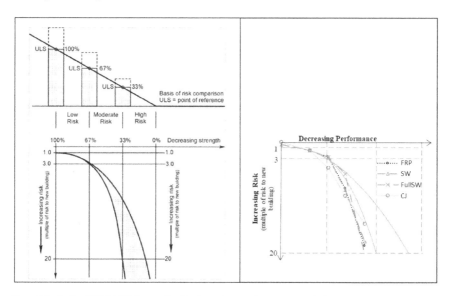

Fig. 1.2 Left: NZSEE2017 - Qualitative (approximate) correlation between the %NBS (Capacity/Demand ratio) and the Relative Seismic Risk (Life Safety) compared to a new building (or a building retrofitted to a %NBS). Right : example of risk reduction following the implementation of alternative retrofit strategies (Fiber Reinfroced Polymers, FRP; Selecrive Weakening, SW; Full Selective Weakening, FW; Column Jacketing, CJ; from (Ligabue et al. 2015).

the seismic response of building designed according to obsolete knowledge/codes and poor details.

Such considerations can suggest interesting corollaries—as confirmed by recent numerical studies (Ligabue et al. 2015, Fig. 1.2 right)—in terms of seismic retrofit strategies. Starting from a high risk building with a %NBS below 33% and a seismic rating D or E, small but focused local retrofit interventions, able to moderately improve the %NBS within the same class, could lead to a substantial and proportionally more significant reduction of the seismic risk in terms of Life Safety, whilst improving the overall reliability of the seismic response of the building, i.e. reducing the dispersion and uncertainties related to the behaviour of the structures.

1.2.2 The New National Plan for Seismic Risk Reduction in New Zealand

In the aftermath of the Canterbury Earthquake Sequence and following the recommendation from the 'Canterbury Earthquake Royal Commission of inquiry' (CERC 2012), the New Zealand Government decided to radically 'change gear' in terms of seismic risk reduction strategies and policies, moving from a 'passive' approach to an 'active' and mandatory one at a national level.

The Building (Earthquake Prone Buildings) Amendment Act 2016 (2016), taking effects from 1 July 2017, introduced major step-changes to the way earthquake prone buildings are identified and managed under the Building Act, under a new national plan for managing buildings in New Zealand. Explicit intent of this national policy framework is to ensure a more consistent and homogeneous approach across the country in the management of the seismic risk for private and public buildings, trying to achieve a balance between (1) Life safety, (2) Cost of strengthening/retrofitting or removing buildings and (3) Considerations on Heritage Buildings.

The new Building Act categorises New Zealand into three seismic risk áreas— high, medium and low—corresponding to peak ground acceleration, PGA, or Z (seismic coefficient in accordance to the NZS1170.5 Loading Standard (2004) of PGA < 0.15; 0.15 ≤ PGA < 0.3; PGA ≥ 0.30, respectively, as shown in Fig. 1.3. Timeframes are set for each seismic areas for identifying earthquake-prone buildings (seismic assessment, within 5–10-15 years respectively) and taking action on them (retrofit or demolition, 15–25-25 years, respectively). A special category of 'priority buildings' is introduced, in high and medium seismic areas, to recognize buildings that are considered higher risk because of their construction, type, use or location. For this category, the timeframe for both assessment and, if need be, remediation is half the time allowed for other buildings in the area.

Important to highlight that the owners should bear the costs for these seismic evaluation and remediation actions. No specific financial incentives, as for example implemented in the recent Italian Guidelines for the classification of the seismic risk (Giovinazzi and Pampanin 2017), were introduced at that stage nor were envisaged in the near future for ordinary buildings in New Zealand. Discussions are taking place in some local municipalities to address the issue of preservation of heritage buildings, often owned by private people, but with a significant yet non-tangible value for the overall community.

Seismic risk area	TAs must identify potentially earthquake-prone buildings within:		Owners must strengthen or demolish earthquake-prone buildings within:	
	Priority	Other	Priority	Other
High	2 ½ years	5 years	7 ½ years	15 years
Medium	5 years	10 years	12 ½ years	25 years
Low	n/a	15 years	n/a	35 years

Fig. 1.3 National risk reduction plan in New Zealand (taking effect from 1 July 2017): timeframe for the identification and the remediation (strengthening/retrofit or demolition) of Earthquake Prone Buildings depending on the level of seismicity (low-medium–high seismic areas—PGA < 0.15 g; 0.15 g ≤ PGA < 0.3 g; PGA ≥ 0.30 g)—MBIE 2016

1.2.2.1   The NZSEE2017 Guidelines on Seismic Assessment of Existing Buildings

As part of the new national risk reduction plan, the Ministry for Business, Innovation and Economy (MBIE) was in charge of the development of new guidelines for the Seismic Assessment (and Rating) of Existing Buildings, with particular focus on a consistent procedure to be used on a daily based by practitioner engineer and Territorial Authorities (TAs) to identify Earthquake Prone Buildings. Following a major multi-year effort of the selected committee, which actually started in 2014 following the CERC Recommendations, the older version of the NZSEE2006 was significantly rewritten and integrated with the more recent lessons and know-how from earthquake events, as well as from the national and international state-of-art and latest research findings.

As it stands, the new NZSEE2017 guidelines "The Seismic Assessment of Existing Buildings" represents the key tool for the actual implementation of the national risk reduction plan. An overview of the motivations, scope and method-ology can be found in (Pampanin 2017), with design example and numerical studies on different Structural systems in (Vecchio et al. 2017, 2018) and (Gentile et al. 2019a, b, c, d), respectively.

In particular, the NZSEE2017 guidelines place significant effort and focus to the use of a simplified analytical procedure—referred to as SLaMA Approach, acronym for 'Simple Lateral Mechanism Analysis', for the evaluation of the capacities (in terms of combination of forces and displacements) of structural elements, connec-tions, subassemblies and seismic-resistant systems, as a fundamental starting point for any vulnerability/safety study and thus prior to any numerical modelling or more sophisticated (but not necessarily more accurate) analysis.

It is an explicit approach to revamp the importance of looking at the big picture with an holistic view, of understanding and controlling the design and thus restore the crucial role of the structural/geotechnical engineers, today often and unfortunately relegated, at least in the views of non-technical customers and decision makers, to the mere role of 'structural calculators'.

Furthermore, as the SLaMa approach allows to highlight the critical struc-tural weaknesses, to evaluate the hierarchy of strength and the sequence of events/mechanisms, it naturally forms the fundamental first step ('diagnosis') to define appropriate retrofit strategies and technique ('prognosis' and 'therapeutic protocol') and critically compare alternative options by considering the overall performance improvement under different earthquake intensity levels, and the achiev-able reduction of the direct and indirect losses, evaluated through Expected Annual Losses (EAL) or similar indicators.

1.2.3 The New Italian Guidelines 2017 Seismic Risk Classification and Financial Incentives

While Life Safety is, and must remain, the highest priority and target of seismic risk reduction policies, the crucial need to consider, assess and, most importantly, reduce the significant direct and indirect losses associated to the damage of structures and infrastructures has become more and more evident. At international level, such considerations have been recently implemented into a regulatory proactive approach for seismic risk reduction based on financial incentives in Italy. Following the significant socio-economic impacts suffered by Italy in the past 20 years following a series of events dislocated across the whole nation (Umbria-Marche 1997; Molise 2002; L'Aquila 2009; Emilia 2012, Central Italy 2016), the government has recently issued (effective from 1 March 2017) a new law providing financial incentives to seismically retrofit privately owned buildings with tax deductions (recoverable in five years, but with the possibility of transferring this tax credit to third parties such as financial institutes, contractors etc.) of 75–85% of the cost of the retrofit intervention) depending on level of improvement in terms of Risk Class (grades A-G) from the status quo (ITA 2017).

Worth noting that the seismic Risk Classification of the ITA2017 guidelines is based on the combination of:

(1)   a Safety Index *IS-V,* (equivalent to the %NBS used in the NZSEE2006 and now NZSEE2017 guidelines) defined as the capacity/demand ratio at Life Safety limit state and

(2)   the Expected Annual Losses, *EAL* (Perdita Annua Media, *PAM*, in Italian) able to take into account the various performances under different intensity levels and limit states

The Risk Class for a building under analysis will be identified as the minimum of the risk classes determined based on the IS-V (=%NBS) and on the PAM (=EAL) assessed grades, i.e. *Risk Class* = MIN (IS-V/%NBS Class; PAM/EAL Class).

Tables 1.1 and 1.2 illustrate the two Risk Class and Grades matrixes according to the NZSEE2017 and ITA2017 Guidelines, respectively. It is worth noting that the

Table 1.1 NZSEE2017 Matrix—% NBS, Grade, Level of Risk, Remedial Actions (modified after per Building Act 2016)

%NBS	Grade	Life risk	Safety prescription
>100	A^+_{NBS}	Low	Acceptable[1]
80–100	A_{NBS}	Low	Acceptable[1]
67–79	B_{NBS}	Low/medium	Acceptable[1]
34–66	C_{NBS}	Medium	Acceptable[2]
20–34	D_{NBS}	High	Legally
<20	E_{NBS}	Very high	Unacceptable[3]

[1]Improvement desirable;   [2]Improvement recommended;   [3]Improvement required

Table 1.2 ITA2017 Matrix—safety index IS-V (=% NBS), risk classes, grades, EAL classes and ranges

IS-V class	IS-V ranges	EAL class	EAL ranges
A^+_{IS-V}	IS-V > 100	A^+_{PAM}	EAL ≤ 0.5%
A_{IS-V}	80% ≤ IS-V < 100%	A_{PAM}	0.5% < EAL ≤ 1.0%
B_{IS-V}	60% ≤ IS-V < 80%	B_{PAM}	1.0% < EAL ≤ 1.5%
C_{IS-V}	45% ≤ IS-V < 60%	C_{PAM}	1.5% < EAL ≤ 2.5%
D_{IS-V}	30% ≤ IS-V < 45%	D_{PAM}	2.5% < EAL ≤ 3.5%
E_{IS-V}	15% ≤ IS-V < 30%	E_{PAM}	3.5% < EAL ≤ 4.5%
F_{IS-V}	IS-V < 15%	F_{PAM}	4.5% < EAL ≤ 7.5%
–	–	G_{PAM}	7.5% ≤ EAL

ranges of %NBS or IS-V adopted to define the Grades or Risk Class are very similar. On the lower classes C-D-E–F (poorer performance) the ITA2017 guidelines have a more detailed classification (and one more class G) while NZSEE2017 code tends to use larger ranges, in recognition of the increasing level of uncertainties associated to the calculation of such index for older and poorly detailed structures.

1.3 The SLaMA Analytical-Mechanical Assessment Procedure

In line with the Park, Paulay and Priestley 'School', in the new NZSEE2017 guidelines, and specifically at the Detailed Seismic Assessment (DSA) level (Tier 2), particular attention has been given to the development of an advanced, reliable while simplified methodology, referred to as SLaMA (Simple Lateral Mechanism Analysis), based on an analytical and mechanical approach, i.e. basically "by hand" or using a spreadsheet, rather than and prior to a numerical (i.e. finite element and computer-based).

The SLaMA procedure for the assessment of the seismic vulnerability/safety and seismic rating (Fig. 1.4) develops from the identifications of the expected critical structural weaknesses through the evaluation of:

- Step 1(a): the flexural and shear capacity (in terms of both forces and displacements) of the structural elements;
- Step 1(b): the hierarchy of strength and sequence of events of the connections and beam-column subassemblies [according to Pampanin and Bolognini (2007)]
- Step 1(c): the local and global collapse mechanisms of the seismic-resisting systems and, finally, the global capacity curve

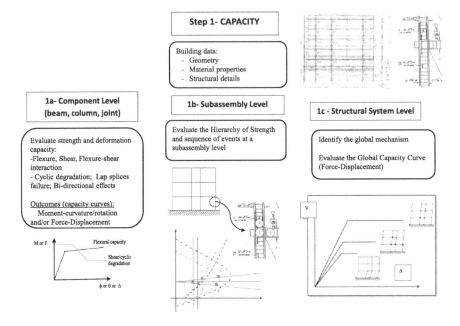

Fig. 1.4 Key steps of the SLaMa analytical procedure according to the NZSEE2017 guidelines (Pampanin 2017) to derive the capacity curve of a building: **1(a)** evaluation of the strength and deformation capacity for flexure and shear of structural components; **1(b)** evaluation of the hierarchy of strength and sequence of events in the beam-column systems and **1(c)** identification of the global mechanism and evaluation of the global force-displacement capacity curve

By comparing the Capacity curve of the structure (in terms of analytical force–displacement non-linear curve) and the Demand (in terms of acceleration-displacement response spectra, ADRS, in line with the capacity spectrum method or similar approaches), the expected performance of the building under different level of shaking intensity, i.e. earthquake return periods, can be evaluated with a relatively simple approach and good level of approximations (Fig. 1.5).

More specifically, given the NZSEE2017 key focus on Life Safety, such an analytical approach allows to evaluate the level of 'relative' (Life) Safety when compared to a newly built structure, by means of the %NBS index (New Building Standard, a capacity-demand ratio) and to associate a seismic risk rating or class (from A to E, with indicative ranges of collapse probabilities when compared to a brand-new structure).

As demonstrated in a series of analytical–numerical comparisons (Bianchi et al. 2019; Gentile et al. 2019a, b, c, d; Vecchio et al. 2017, 2018), the SLaMa analytical procedure provides quite reliable results, particularly satisfactory when considering the simplicity of the method.

Refinement of the results can be obtained by preparing a non-linear numerical lumped plasticity model, 'informed' by the hierarchy of strength and local mechanism predicted by SLaMA, and running pushover or time history analyses.

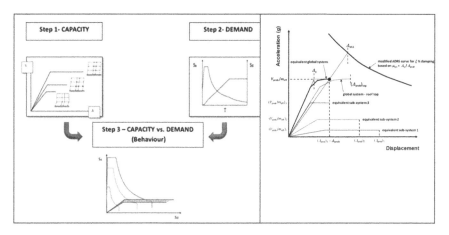

Fig. 1.5 Left: Evaluation of the seismic performance at different level of intensity by intersecting the Capacity (force–displacement) curve with the Demand (ADRS—Acceleration-Displacement Response Spectra). Right: evaluation of the %NBS as Capacity/Demand ratio, (modified after NZSEE2006–2017)

1.3.1 Selection of Retrofit Strategies and Techniques

The SLaMa method can be used for a relatively quick yet reliable – when compared to what required by more complex and detailed non-linear pushover and time-history analyses – estimation of the expected behavior and performance of the building (or classes of buildings) before and after a retrofit/strengthening intervention, thus becoming a fundamental supporting tool for the implementation of a medium-long term strategy of seismic-risk reduction at national scale.

In fact, as part of the analytical evaluation of the force-displacement capacity curve of the system, the sequence of local and global mechanisms can be captured (i.e. what happens at what stage). More specifically, as shown in Fig. 1.6, the values of chord rotations, interstorey drifts, (top floor or effective height) displacements corresponding to each event (as the achievement of ULS shear or flexure in a beam, column or joint) can be identified and clearly visualized in a Force–Displacement or ADRS domain.

1.3.2 Quantifications of Impairment—Loss Estimation

Following a capacity spectrum approach and thus intersecting the capacity curves with the various level of earthquake (shaking) intensities (and thus return Period T_r), it is possible to evaluate what intensity level (expressed in Peak Ground Acceleration, PGA) and thus what associated Return Period (T_r) would correspond to the achievement of the four level of performances/damage/limit states, namely, i.e. Immediate

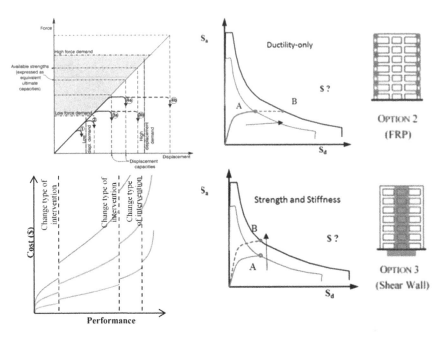

Fig. 1.6 Top Left: use of SLaMA method to capture the sequence of events/mechanisms within an analytically-derived pushover capacity curve. Right: selection of alternative retrofit strategies and techniques to achieve the targeted performance (Ligabue et al. 2015). Bottom Left: qualitative ladder-scale representing the expected cost of retrofit intervention as a function of the targeted performance or %NBS (adapted from (Beetham 2013))

Occupancy (IO), Damage Control (DC), Life Safety (LS) and Collapse Prevention (CP), as shown in Fig. 1.7.

The relationship between PGA and Return Period depends on the local seismicity and earthquake magnitude.

The Italian Guidelines ITA2017 provide a relationship and a power coefficient (average value used herein, in principle it is function of the local seismicity and magnitude) to convert PGA into T_r and vice versa, as follows:

$$T_{rC} = T_{rD}(PGA_C/PGA_D)^{1/0.41} \tag{1.1}$$

The sub indexes C and D indicate 'Capacity' and 'Demand' respectively.

In order to create a MAF versus % Reconstruction Cost curve it is necessary to estimate the relationship between an Engineering Demand Parameter (e.g. Drift, acceleration), the Level of Damage/Limit States and the repairing costs associated to that level of damage.

Following the L'Aquila 2009 earthquake in Italy and the repair/reconstruction phases, covered by government funds through a very comprehensive scheme, a unique and detailed database of costs of reconstruction versus damage levels was

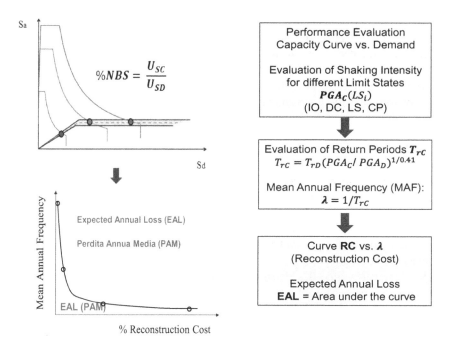

Fig. 1.7 Left: Procedure for the evaluation of the Expected Annual Loss (EAL): (1) Comparison of Capacity Curve and Demand (ADRS Spectra); (2) Evaluation of the earthquake intensity levels PGAc and associated return period Trc, corresponding to the various Limits States (IO, DC, LS, CP). Note that in the top left-side graph only three Limit States are herein for simplicity indicated, namely IO < DC < LS); (3) Plot the curve Mean Annual Frequency (MAF or $\lambda = 1/T_{rC}$) versus Costs (% reconstruction costs); (4) Evaluation of the EAL as integral (area) of this curve

collected and published in a "White Paper" or "Libro bianco" (Ludovico et al. 2017a, b).

Based on these data, a table correlating the expected (mean values of) % Reconstruction Costs corresponding to a given limit state/observed damage was developed. As shown in Table 1.3, such correlation is a key part of the simplified procedure to evaluate the Expected Annual Loss (EAL) starting from a capacity versus demand comparison as described above.

Table 1.3 Correlation between the various Limit States (IO, DC, LS, CP) and the Reconstruction Costs according to the ITA2017 guidelines

Limit states	Reconstruction cost RC (%)
Collapse or total economic loss	100
Near collapse, NC	80
Life safety, LS	50
Damage control, DC	15
Immediate operational, IO	7
Zero loss	0

The Expected Annual Loss, evaluated as the integral of the MAF versus $1/T_r$ curve thus the Area underneath that curve, represents the overall repairing costs so calculated/estimated during the life-span of the building divided by the 50 years (or whatever assumed as a Life-span) of the structure.

This comprehensive procedure to evaluate the Expected Annual Loss (EAL) directly from capacity curves and earthquake demands (ADRS Spectra) is well recognized within the scientific community but unique at a worldwide level in terms of regulatory provision.

Such an approach is a direct and practical response to the lessons learnt from recent earthquakes, acknowledging and highlighting the severe and significant repairing costs associated to non-structural elements (partitions/ infills, facades, ceilings, services) under "frequent' earthquakes (i.e. minor-moderate earthquake and relatively small return period T_r of 25–100 years), when compared to more 'rare' events (i.e. major-severe intensity and large return period T_r of 500–1000 years or above) typically used as a design level reference, when thus focusing on "Life Safety" as primary Limit State.

In addition to a Time-Based loss estimation of the EAL/PAM, following a similar approach, a Scenario-Based Assessment can be used to estimate the expected level of direct economic loss, again expressed in terms of % Reconstruction Costs under a specific event or scenario earthquake, such as:

(1)   the Probable Frequent Loss (PFL) defined by Porter et al. (2004a) as the mean loss resulting from shaking with a 10% exceedance probability in 5 years (i.e. Return Period Tr = 50 yrs) and/or

(2)   the Probable Maximum Loss defined as the mean loss resulting from shaking with a 10% exceedance probability in 50 years (i.e. Tr = 500 yrs)

This latter Scenario-Based Assessment approach focuses into a specific window within the building life-time response and provides a complementary information to the EAL, which instead gives a more comprehensive, yet general and thus spread across the various limits states (and earthquake intensity), view of the overall building performance under a range of seismic excitations.

1.4   Post-Earthquake Residual Capacity of Damaged Buildings

In New Zealand, the Canterbury Earthquake Sequence 2010–2011 and, again few years later, the Kaikoura Earthquake 2016, similarly to the several earthquake events occurred in the past decades at international level, have highlighted the complexity and uncertainties associated with the delicate decision of repair versus demolition, due to the lack of: (a) evidence-based information/knowledge, (b) codified ad-hoc guidelines related to the (b1) evaluation of the residual capacity of a damaged building to sustain subsequent aftershocks and (b2) the selection and implementation of appropriate cost-effective repairing techniques to 'restore' the building to its pre-earthquake conditions (CERC 2012).

In the past years a significant research effort has been dedicated at both national and international level [US, Europe and Japan, FEMA 306 (1998), JBDPA (2015)] to gain a better understanding on the residual capacity of buildings following an earthquake event (Cuevas et al. 2014, 2015; Cuevase and Pampanin 2016, 2017; Loporcaro et al. 2014. 2017, 2018, 2019; Luco et al. 2004; Ludovico et al. 2013; Maeda 2008; Maeda and Eon-Kang 2009; Maeda et al. 2012, 2017; Mukai et al. 2017; Polese et al. 2012).

In New Zealand, the on-going findings and key outcomes of the first NZ-based project on Residual Capacity and Repairing Options [NHRP Funded, Contestable round 2012, (Pampanin et al. 2015)], have formed the basic knowledge platform of the activities of a special MBIE (Ministry of Business, Innovation and Employment) committee/working group on the topic residual capacity (Cuevas et al. 2014, 2015; Cuevas and Pampanin 2016, 2017; Loporcaro et al. 2014, 2017, 2018, 2019), leading to the development of a draft framework for the evaluation of the residual capacity of buildings (Elwood et al. 2016).

Further developments have been obtained in the recent past, suggesting to merge the benefits of different methodologies into one general analytical-mechnical approach, based on an evolution of the SLaMa method, as described and proposed in the following sections.

1.4.1 Effects of Damage on Future Building Performance

Provided appropriate/reliable reduced capacity curves are derived, the aforementioned general SLaMA-based approach for the assessment of seismic safety as well as the estimation of losses can be implemented for either the as-built (un-damaged), the damaged and the repaired/retrofitted (with different solutions and targeted performance) configurations, providing valuable insights to support the decision making process for various stakeholders (e.g. alternative retrofit solutions, territorial scale approach, repair versus demolition, cost/benefit analysis).

1.4.1.1 General Procedure to Evaluate the Impairment of the Capacity Curves (FEMA306/307-Type)

"Understanding the effects of damage on future building performance" has been the declared first 'objective' of the ATC-43 project (which delivered the FEMA306 and 307 guidelines) building on a thorough review (yet dated 1998) of available analysis techniques, field observations, test data, and emerging evaluation and design methodologies, and complemented by further analytical/numerical work.

As noted in the FEMA 306 document *"The quantitative evaluation of the effects of earthquake damage on structures requires the selection of a measurement parameter. Procedures in this document use change in the anticipated performance of the building during future earthquakes as the measurement parameter. This is the change due directly to effects of earthquake damage on the basic structural properties that control seismic performance. If the structural property changes are estimated, the corresponding change in future performance can also be estimated".*

The quantitative procedure proposed to assess and compare the relative performance of the building in its pre-damage, post-damage and repaired state relies upon non-linear static analysis techniques (i.e. capacity spectrum method), thus fully consistent with, and achievable with, the SLAMa Method.

The Capacity Curves of the overall structures can be derived for the as-built, the damaged or repaired level, starting from the capacity curve of a component level (see Fig. 1.8) and depending on:

(a)   component type (wall, pier, spandrel, coupling beam);
(b)   extent of damage (minor-moderate-severe) and
(c)   mode of damage/behaviour (ductile vs. brittle)

The effects of damage on the component backbone curve behaviour are taken into account by means of modification/reduction factors of stiffness (λK), strength (λQ), deformation capacity (λD) and Residual Drift (RD), calibrated numerically and experimentally.

The values of these modification/reduction factors, for each component of the structural system, depend on the type of damage, according to the behaviour mode (ductile vs. brittle, flexure vs. shear prone) and the severity of damage (insignificant, slight, moderate, severe, extreme).

The FEMA 306 and 307 Guidelines provides a component Damage Classification Guide in the form of table/charts/schetces to assist the engineer with the classification of type (behaviour mode) and severity of the damage, and derive associated component modifications factors. Modifications factors are also presented to account

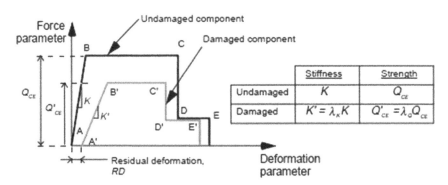

Fig. 1.8 Component modelling criteria and stiffness/strength/displacement reduction factors to account for damage (FEMA 306 (1998))

for repair/restoration. The documents primarily focuses on reinforced concrete and masonry wall buildings, with some insights into infills and their interactions with the surrounding frames. Limited information is available in literature for other type of structural systems and components, such as beam-column joints, beam plastic hinges, column flexural, lap splice or shear failure, floor/diaphragm behavior. Yet, recent numerical/experimental investigation in the past years have started filling the gap: in (Ludovico et al. 2013; Polese et al. 2012) modification factors were proposed to evaluate the residual capacity of non-confirming RC columns typical of the Mediterranean region; furthermore, experimental and numerical investigations on cumulating damage, residual capacity and repairing options for plastic hinges in beam-column subassemblies and beam cantilever as well as on the effects on strain hardening and low-cycle fatigue in steel rebars on the overall plastic hinge behavior and repairability have been carried out in New Zealand (Cuevas et al. 2014, 2015; Cuevas and Pampanin 2016, 2017; Elwood et al. 2016; Loporcaro et al. 2014, 2017, 2018, 2019; Luco et al. 2004; Maeda 2008; Maeda and Eon-Kang 2009; Maeda et al. 2012, 2017; Mukai et al. 2017; Pampanin et al. 2015).

1.4.1.2 Proposed SLaMA-Based Procedure for Pre- and Post-Earthquake Safety Evaluation and Loss Assessment of Buildings

Building on, a merging, the aforementioned recent development and approaches, and relying upon the current modifications factors and/or deriving specific ones (based on numerical/experimental evidences) as needed, a general and consolidated SLaMA-based method can be adopted, moving from component level to subassembly and structural system level, for the safety evaluation and loss assessment of buildings in either pre- or post-earthquake, in the as-built or retrofitted/repaired, configurations.

Figure 1.9 illustrated the steps and qualitative outcome of such loss estimation procedure when a building structure is subject to a reduction of stiffness and/or strength: for three different levels of damage, with associated levels of stiffness λK and strength reduction, λQ, the area underneath the Mean Annual Frequency and the % Replacement Costs will increase, leading to three different level of Expected Annual Loss, EAL.

In general terms, depending on the assessed level of damage occurred in each component and subassembly (to be evaluated through a SLaMA force–displacement curve and ADRS spectrum), the capacity curves (momento-rotation, force–displacement) of each component can be updated with the appropriate corresponding modifications factors (in terms of stiffness, strength, deformation capacity and residual drfit/chord rotation) as provided by the most current and relevant literature. A second iteration of the SlaMA methology on the building, assuming this damage state configuration, would identify new critical structural weaknesses and possible new sequence of events, due to a modified internal hierarchy of strength, thus leading to local and global mechanism. ADRS curves would be selected to represent either the code design level, whereby a post-earthquake safety check is required, or selected levels

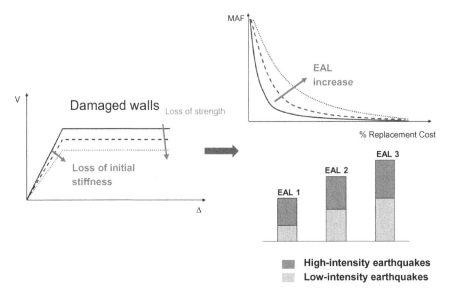

Fig. 1.9 Expected impact on performance and EAL due to loss of stiffness and/or strength as it can be evaluated following a SLaMA-based analytical-mechanical procedure

of expected aftershocks or scenerio-based events. The effects of cumulating cycles can be implemented in the procedure by means of strength degradation curves, either depending on the number of cycles and/or level of ductility at a member level, as well as bond-deterioration or stress-decay curve at a material level.

In addition to safety check and residual capacity (to withstand further aftershocks or new events) considerations, various type of losses, as previously described, e.g. EAL, PFL, or PML or other KPIs (Key Performance Indicators) such as the loss of building operability (downtime), can be evaluated, providing very valuable support to the decision making process.

As an example, it could be assessed, as done in a previous work by (Beetham 2013), that the Probable Frequent loss (PFL) is directly (possibly but not necessarily linearly) proportional to the (loss of) initial stiffness or fundamental period of vibration of the structure, as shown in Fig. 1.10 for the specific case of an existing frame building with or without alternative retrofit techniques.

Such information, related to the potential value, or loss of value, of a building in the aftermath of minor-moderate earthquake events (or in the case of aftershocks following the main shock) could be of direct interest and practical use as part of the discussion between insurance/reinsurance companies and owners or other interested parties.

Similarly, alternative retrofit or repair/retrofit strategies can be compared not only in terms of Safety Index improvement (%NBS or IS-V) but also in terms of economical (losses) considerations as shown in Figs. 1.11 and 1.12, allowing for a more

Fig. 1.10 Trend of Probable Frequent Loss (PFL) as a function of the % Building Replacement Cost versus Fundamental Period of the Structures (or loss of stiffness) (Beetham 2013)

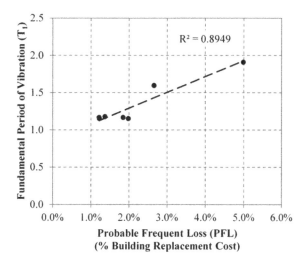

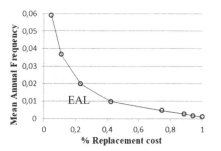
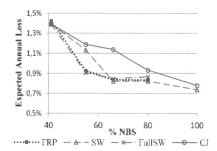

Fig. 1.11 Left: Example of Mean Annual Frequency, MAF, versus % Replacement cost: the area under the curve represents the Expected Annual Loss, EAL (Beetham 2013); Right: Values of EAL for different retrofit strategy and % NBS achieved (Ligabue et al. 2015)

informed decision based on (possibly and preferably multi-criteria) cost–benefit evaluation on whether or not, and to what level, it would be worth pursuing a specific retrofit/repair intervention program.

The methodology could also allow to investigate the opportunity to implement hybrid intervention at either individual building level or assets of buildings at territorial scale level, with the negotiated involvement of private, public and insurance sectors. As an example, the opted strategy could consist of a partial retrofit intervention, beared partially by the owner, as part of a self-insurance approach, and/or partially supported by direct financial incentives from the government, and/or in combination with an insurance premium rebate, adequately adjusted to recognize the enhanced and quantifiable level of protection achievable both in terms of Life Safety as well as Asset/building/Business as result of the retrofit intervention (Fig. 1.13).

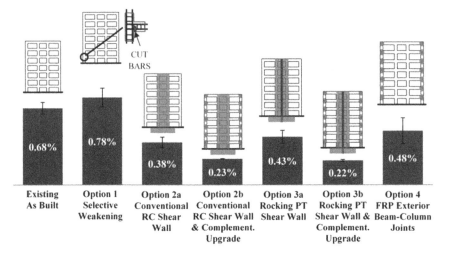

Fig. 1.12 Comparison of alternative retrofit solutions in terms of Expected annual loss (EAL) as a percentage of the building replacement cost (Beetham 2013)

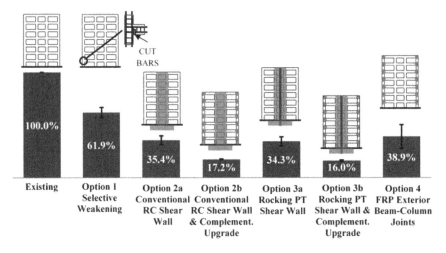

Fig. 1.13 Comparison of alternative retrofit solutions in terms of Probable maximum loss (PML)—under a Design Level earthquake (500 years return period or 10% probability of exceedance in 50 years) (Beetham 2013)

1.5 Concluding Remarks

This chapter has provided an overview of recent developments at international level in terms of methodology and regulatory approaches as well as "diagnosis" protocols and procedures for either pre- and post-damage of existing buildings.

Building on, and merging, these recent developments, an integrated methodology for the evaluation and assessment of either pre-earthquake as well as post-earthquake safety evaluation and loss assessment of buildings, based on a simplified analytical-mechanical approach, referrred to as SLaMA (Simple Lateral Mechanism Analysis), has been presented, in order to support the engineering community as well as the various stakeholders through the various steps of the decision making process of risk assessment and reduction.

Worth reminding that, in addition to the opportunity and value of sharing the technical content of these recent developments, the key goal remains to stimulate awareness, discussion within, as well as outside (equally if not more important) the technical community, and foster synergetic political actions for a more proactive and comprehensive implementation of seismic risk reduction programs and international level.

References

Building Act (2004) Department of building and housing-Te Tari Kaupapa Whare, ministry of economic development, New Zealand Government, Wellington, New Zealand

Beetham T (2013) Simplified displacement based performance assessment of retrofit alternatives for decision making. Master thesis, IUSS, University of Pavia, Italy

Bianchi S, Ciurlanti, J, Pampanin S (2019) A SLaMA-based analytical procedure for the Cost/Performance-based evaluation of buildings. In: Compdyn Conference, Crete, Greece

Building (Earthquake-prone Buildings) Amendment Act (2016) New Zealand Government, Wellington, New Zealand

CERC (2012) Canterbury Earthquake Royal Commission, https://canterbury.royalcommission.gov t.nz

Cuevas RA, Pampanin S (2016) Assessing the seismic residual fatigue life of reinforced concrete frame buildings: a proposed framework. In: 16th U.S.-Japan-New Zealand Workshop on the Improvement of Structural Engineering and Resiliency, Nara, Japan

Cuevas A, Pampanin S (2017) Post-seismic capacity of damaged and repaired reinforced concrete plastic hinges extracted from a real building. In: 16th WCEE, World Conference on Earthquake Engineering, Santiago, Chile, January 9–13, 2017 paper 3142

Cuevas A, Akguzel U, Pampanin S (2014) Preliminary numerical investigation on the seismic residual capacity of reinforced concrete plastic hinges. In: 2nd European Conference in Earthquake Engineering and Seismology, Istanbul 25–29 Aug

Cuevas A, Pampanin S, Carr A, Ozbolt J (2015) Seismic residual capacity of reinforced concrete frames: Part I: General Framework. Research Report UC 2015–4, Part I, Department of Civil and Natural Resources Engineering, University of Canterbury, Christchurch, New Zealand

Del Vecchio C, Gentile R, Pampanin S (2017) The simple lateral mechanism analysis (SLaMA) for the seismic performance assessment of a case study building damaged in the 2011 Christchurch Earthquake. University of Canterbury Department of Civil and Natural Resources Engineering, Research Report 2016–02, New Zealand, ISSN 1172–9511

Del Vecchio C, Gentile R, Di Ludovico M, Uva G, Pampanin S (2018) Implementation and validation of the simple lateral mechanism analysis (SLaMA) for the seismic performance assessment of a damaged case study building. J Earthq Eng Doi/full/. https://doi.org/10.1080/13632469.2018.1483278

Di Ludovico M, Polese M, Gaetani d'Aragona M, Prota A, Manfredi G (2013) A proposal for plastic hinges modification factors for damaged RC columns. Eng Struct 51:99–112

Di Ludovico M, Prota A, Moroni C, Manfredi G, Dolce M (2017a) Reconstruction process of damaged residential buildings outside historical centres after the L'Aquila earthquake: part I—" light damage" reconstruction. Bull Earthq Eng 15(2):667–692

Di Ludovico M, Prota A, Moroni C, Manfredi G, Dolce M (2017b) Reconstruction process of damaged residential buildings outside historical centres after the L'Aquila earthquake: part II-"heavy damage" reconstruction. Bull Earthq Eng 15(2):693–729

Elwood KJ, Marder K, Pampanin S, Cuevas RA, Smith P, Cattanach A, Stannard M (2016) Draft framework for assessing residual capacity of earthquake-damaged concrete buildings. In: New Zealand Society of Earthquake Engineering Conference, Christchurch, New Zealand

FEMA 306 (1998) Evaluation of earthquake damaged concrete and masonry wall buildings: Basic procedures manual. ATC, Redwood City, CA, USA

Gentile R, Del Vecchio C, Pampanin S, Raffaele D, Uva G (2019a) Refinement and validation of the simple lateral mechanism analysis (SLaMA) procedure for RC frames. J Earthq Eng. https://doi.org/10.1080/13632469.2018.1560377

Gentile R, Pampanin S, Raffaele D, Uva G (2019b) Non-linear analysis of RC masonry-infilled frames using the SLaMA method: Part 1—mechanical interpretation of the infill/frame interaction and formulation of the procedure. Bull Earthq Eng. https://doi.org/10.1007/s10518-019-00580-w

Gentile R, Pampanin S, Raffaele D, Uva G (2019c) Non-linear analysis of RC masonry-infilled frames using the SLaMA method: Part 2—parametric analysis and validation of the procedure. Bull Earthq Eng. https://doi.org/10.1007/s10518-019-00584-6

Gentile R, Pampanin S, Raffaele D, Uva G (2019d) Analytical seismic assessment of RC dual wall/frame systems using SLaMA: Proposal and validation. Eng Struct 188:493–505. https://doi.org/10.1016/j.engstruct.2019.03.029

Giovinazzi S, Pampanin S (2017) Simplified approaches for the seismic risk rating of reinforced concrete buildings and the selection of retrofit strategies. In: XVII ANIDIS, Italian Association for Earthquake Engineering, Pistoia, 17–22 Sept

ITA2017 (2017) Annex A to the D.M. n. 65 del 7–03–2017 "Linee Guida per la Classificazione del Rischio Sismico delle Costruzioni" (in Italian)

JBDPA (2015) Guideline for Post-earthquake Damage Evaluation and Rehabilitation (in Japanese), Japan Building Disaster Prevention Association

Ligabue V, Savoia M, Pampanin S (2015) Repairing/retrofitting vs. replacing? evaluating the cost-effectiveness of alternative options to support decision making, NZSEE Conference, Rotorua

Loporcaro G, Pampanin S, Kral MV (2014) Investigating the relationship between hardness and plastic strain in reinforcing steel bars. In: NZSEE Conference, Auckland, 21–23 March, Paper No. P22

Loporcaro G, Kral M, Pampanin S (2017) Case study: application of the hardness method to estimate the residual capacity of reinforcement in an earthquake damaged building. In: 16th World Conference on Earthquake Engineering, Santiago, Chile, January 9–13, 2017, paper n. 2075

Loporcaro G, Pampanin S, Kral MV (2018) Estimating plastic strain and residual strain capacity of earthquake-damaged steel reinforcing bars. J Struct Eng 144(5). https://doi.org/https://doi.org/10.1061/(ASCE)ST.1943-541X.0001982

Loporcaro G, Pampanin S, Kral MV (2019) Long-term strain-ageing effects on low-carbon steel reinforcement. Const Build Mater 228. https://doi.org/https://doi.org/10.1016/j.conbuildmat.2019.07.332

Luco N, Bazzurro P, Cornell CA (2004) Dynamic versus static computation of the residual capacity of a mainshock-damaged building to withstand an aftershock. In: 13th World Conference on Earthquake Engineering, Vancouver, Canada. Paper no. 2405

Maeda M (2008) Effect of seismic retrofit on seismic risk mitigation for reinforced concrete buildings. In: 14th World Conference on Earthquake Engineering, Beijing, China

Maeda M, Eon-Kang D (2009) Post-earthquake damage evaluation of reinforced concrete buildings. J Adv Conc Tech 7(3):327–335

Maeda M, Nakano Y, Lee KS (2012) Post-earthquake damage evaluation for R/C buildings based on residual seismic capacity. In: Proceedings of the 13th World Conference on Earthquake Engineering, Vancouver, Canada

Maeda M, Nishida T, Matsukawa K, Murakami M (2017) Revision of guideline for postearthquake damage evaluation of reinforced concrete buildings in Japan. In: 16th World Conference on Earthquake Engineering, Santiago, Chile

MBIE (2016)—Building (Earthquake prone Buildings) Amendment Act, 2016, Ministry of Business, Innovation and Employment, NZ. Portal 'Managing Earthquake Prone Buildings', https://www.building.govt.nz/managing-buildings/managing-earthquake-prone-buildings/

Mukai T, Kabeyasawa T, Tani M, Suwada H, Fukuyama H (2017) Residual seismic capacity of ductile RC frame with wing walls based on full-scale loading test. Bull New Zealand Soc Earthq Eng 50(44):565–573

NZS1170.5 (2004) Structural design actions. Part 5: Earthquake actions—New Zealand, Standards New Zealand, Wellington, NZ

NZSEE (2006) Assessment and improvement of the seismic performance of existing buildings, New Zealand Society of Earthquake Engineering, Wellington, https://www.nzsee.org.nz/publications/assessment-and-improvement-of-the-structural-performance-of-buildings-in-earthquake/

NZSEE (2017) The seismic assessment of existing buildings (the guidelines), New Zealand Society of Earthquake Engineering, Wellington, Version October 2016. https://www.eq-assess.org.nz/

Pampanin S (2017) Towards the practical implementation of performance-based assessment and retrofit strategies for RC buildings: challenges and solutions. In: SMAR2017- Fourth conference on Smart Monitoring, Assessment and Rehabilitation of Structures, Keynote Lecture, 13-15 Sept, Zurich, Switzerland

Pampanin S, Bolognini, Pavese D (2007) Performance-based seismic retrofit strategy for existing reinforced concrete frame systems using FRP composites. ASCE J Comp Const, Special Issue on "Recent International Advances in FRP Research and Application in Construction 11(2):211–226. ISSN 1090–0268/2007/2–211–226

Pampanin S, Cuevas A, Kral M, Loporcaro G, Scott A, Malek A (2015) Residual capacity and repairing options for reinforced concrete buildings, Research Report Prepared for the Natural Hazard Research Platform, Contract 2012-UOC-02-NHRP

Polese M, Di Ludovico M, Prota A, Manfredi G (2012) Residual capacity of earthquake damaged buildings. In: Proceedings of the 15th World Conference on Earthquake Engineering, Lisboa, Portugal

Chapter 2
Damage Assessment in Japan and Potential Use of New Technologies in Damage Assessment

K. Kusunoki

Abstract Right after an earthquake, it is quite important to evaluate the damage level of the buildings in the affected area. In Japan, a rapid inspection is conducted to evaluate the risk of collapse due to an aftershock. If any damage is detected, it is required to conduct damage classification, which takes time but categorizes its damage into five damage categories. Japan has a standard for both rapid inspection and damage classification. They are briefed in this chapter. Similar to the damage classification, the loss of the house and home contents for the earthquake insurance. The method for earthquake insurance is also introduced. Since they are based on visual inspection, it is quite difficult to investigate the damage of the high-rise buildings and buildings covered by finishing. Recently, many kinds of research are conducted to use sensors for automatic and realtime damage classification. A structural health monitoring method with accelerometers based on the capacity spectrum method, which is currently installed into more than 40 buildings, is also introduced.

2.1 Introduction

Japan is one of the earthquake-prone countries. We apply a seismic code that requires a very high seismic performance of which base-shear coefficient demand for the short-period building is 1.0. Since the demand is too high to keep the buildings elastic, non-linear behavior such as flexural yielding is accepted to dissipate the input energy safely and to reduce the demand. The base-shear coefficient demand for the most ductile reinforced concrete building is 0.30. It can be said that the buildings may suffer damage during a severe earthquake.

Rapid inspection of existing structures soon after a big earthquake is crucial in order to prevent tragedies due to aftershocks. Civil infrastructures such as public buildings that are supposed to be used as shelters need to be evaluated to find out the seismic performance during aftershocks. On the other hand, it is also very important

K. Kusunoki (✉)
Earthquake Research Institute, The University of Tokyo, 1-1-1, Yayoi, Bunkyo Ku, Tokyo, Japan
e-mail: kusunoki@eri.u-tokyo.ac.jp

© The Author(s) 2021
S. Akkar et al. (eds.), *Advances in Assessment and Modeling of Earthquake Loss*,
Springer Tracts in Civil Engineering,
https://doi.org/10.1007/978-3-030-68813-4_2

to screen out the buildings that still have enough seismic capacity soon after a main-shock, since a lot of people may refuge from their houses due to fear of collapse even if they have enough capacity. It can help reduce the number of refugees.

In this chapter, the rapid inspection method in Japan (Japan building disaster prevention association 2015) is introduced. If any damage is detected, the damage level is classified into six classes, "none", "minor", "slight", "moderate", "severe", and "collapse" according to the more detailed investigation. It is called the damage classification method. The standard is available in Japan to classify the damage of the affected building and to evaluate the capacity if strengthening is needed when it is repaired. The outline of the standard is also introduced.

Same as the rapid inspection, the damage level of the affected building needs to be evaluated right after an earthquake for earthquake insurance. The amount of insurance payment should be paid according to its damage level. The method of the rapid damage assessment for earthquake insurance is also introduced in the paper (The general insurance association of Japan 2019).

Currently, buildings have to be inspected one by one by engineers or researchers according to the above three methods. For example, 5,068 engineers and 19 days were needed to conduct the rapid inspection with 46,000 buildings on a damaged area at the Kobe earthquake. Nineteen days were too long, and yet the number of inspected buildings was not enough. Moreover, many buildings were judged as "Limited entry", which needs a detailed assessment by engineers. "Limited entry" judgment is a gray zone, and it could not take away anxieties from inhabitants. Furthermore, the current rapid inspection system presents a dilemma since buildings should be inspected by visual observation of engineers. Thus, judgment varies according to engineers' experience.

In order to solve the problems mentioned above, the author has been developing the real-time residual seismic capacity evaluation system, which needs only few relatively inexpensive accelerometers. The system calculates the performance and demand curves from a measured acceleration of the basement and of each point of a structure with inexpensive accelerometers, and further estimate the residual seismic capacity of a structure by comparing these curves. To draw the performance curve, the absolute response accelerations, and relative response displacement at each point are needed. The displacements are derived from the accelerations by the double integral in the system. The outline of the system and the result obtained from the recorded data of an instrumented building during the 2011 Tohoku Earthquake will be presented.

2.2 Rapid Inspection Method in Japan

The rapid inspection method and the damage classification method were developed by the project "Development of the restoration techniques for damaged building due to an earthquake" (1981–1985) funded by the ministry of construction. Manual for post-earthquake rehabilitation techniques for buildings (draft) was published in

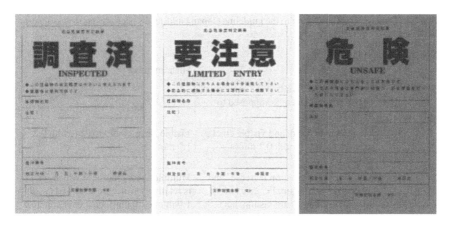

Fig. 2.1 Tags according to the rapid inspection result (Japan building disaster prevention association 2015)

1986. The manual says that the rapid inspection is for evaluating the risk of collapse and falling of nonstructural elements, and for tagging "Unsafe", "Limited Entry", and "Inspected" as shown in Fig. 2.1 (Japan building disaster prevention association 2015).

The rapid inspection method is applied in Japan to "rapidly" figure out risky buildings against consequent aftershocks. The inspection is based on the visual observation from outside of the buildings. The risks of both structure and foundation are assessed accordingly. The damage levels of the structural members are classified into five damage classes according to their crack patterns and their residual crack width. Firstly, each member is categorized as "flexural member" and "shear member" by the damage pattern or construction year. Secondly, the damage of the member is classified according to the residual crack width, category, and damage condition, as shown in Table 2.1. Each damage class is conceptually defined based on the dissipated strain energy E_d and remaining strain energy E_r, as shown in Fig. 2.2.

The damage of the structure is evaluated with Table 2.2. The inspection is conducted for the most damaged story. If the damage of any member is classified as Damage Class III, IV, or V, the building is classified as "Rank B". If the ratio of the number of the columns classified as Damage Class IV and V is high, the building is ranked as B, or C. If the inclination due to uneven settlement is large, it is ranked as B or C. The risk of neighboring buildings and foundation is also taken into account. With all evaluated ranks, the building is categorized as "Unsafe" if there is a Rank C or more. If all risks are evaluated as Rank A, the building is categorized as "Inspected". Other buildings are categorized as "Limited Entry".

The inspector must be 1st or 2nd class licensed architect or timber building architect who is living in the affected area. The inspector needs to take a lecture provided by the local government and to be registered. The rapid inspection is supposed to start soon after an earthquake and to finish within seven days.

Table 2.1 Damage class according to the guideline (Japan building disaster prevention association 2015)

Damage class	Condition	
	Flexural member	Shear member
I	Just fine cracks (width < 0.2 mm) exist, but no reinforcement is supposed to yield	
II	Member may yield, and visible cracks exist at its ends (width 0.2 mm ~ 1.0 mm)	Visible shear cracks exist (width 0.2 ~ 1.0 mm)
III	Non-linear deformation increases and relatively wide flexural cracks (width 1.0 mm ~ 2.0 mm) are visible, but cover concrete does not fall much, and core concrete is sound	Multiple shear cracks, of which width is relatively wide, are observed (width 1.0 mm ~ 2.0 mm), but cover concrete does not fall much, core concrete is sound, and restoring force seems to remain
IV	There are many wide cracks, cover concrete falls a lot, and core concrete gets damaged, and reinforcement is visible. Lateral force carrying capacity may be reduced, but columns and walls still carry the gravity load	There are many wide shear cracks, cover concrete falls a lot, and core concrete gets damaged, but buckling/fracture of rebar or hoops are not observed. Lateral force carrying capacity may be maintained
V	Rebar buckled, and even core concrete falls. It seems almost no lateral load carrying capacity is left. Columns/walls shorten. Inclination or settlement may be observed. Rebar may fracture	

Fig. 2.2 Damage class according to the energy dissipation (Bunno et al. 2006)

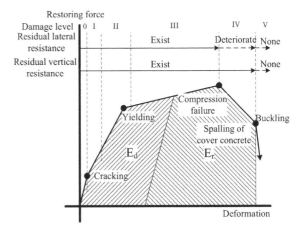

Table 2.2   Rapid inspection result according to the risks of foundation and structure (Japan building disaster prevention association 2015)

		Rank A	Rank B	Rank C		
(1)	Damage level III or more exists	No	Yes	–		
	Neighboring building or foundation looks dangerous	No	Unknown	Yes		
	Inclination due to uneven settlement	<1/60	1/60–1/30	1/30<		
	Damage of column (The most severely damaged floor th floor)					
(2)	Num. of columns with damage level V inspected ratio %			<1%	1–10%	10%<
	Num. of columns with damage level IV inspected ratio %			<10%	10–20%	20%<
	Judgment		Inspected All are rank A	Caution Only one rank B	Unsafe Others	
Overall judgment (Take worse case between (1) and (2)	Inspected		Caution		Unsafe	

The risks of the non-structural elements, window and frame, wet and dry finishing, signboard/machinery, outdoor staircase, and other, are evaluated with Table 2.3. If all risks are evaluated as Rank A, the building is categorized as "Inspected". Other buildings are categorized as "Limited Entry". Finally, the building is tagged with the worse category among structural and non-structural damage categories.

2.3   Damage Classification

Once structural damage is observed, a more detailed assessment is needed to evaluate if the building should be repaired or demolished according to not only the damage level but also the seismic intensity at the site. "The standard to classify the damage level due to an earthquake" is applied for the assessment. Firstly, the seismic index of the building without considering the damage, I_s, is calculated with Eq. (2.1).

Table 2.3 Rapid inspection result according to the risks of non-structural elements (Japan building disaster prevention association 2015)

	Rank A	Rank B	Rank C
Window, frame	Almost no damage	Deformed/cracked	High risk to fall
Wet finishing	Almost no damage	Partial damage	Significant damage
Dry finishing	Fine crack in joints	Gap observed	Significant shift
Signboard/machinery	No inclination	Slight inclination	High risk to fall
Outdoor staircase	No inclination	Slight inclination	Significant inclination
Others ()	Safe	Caution	Unsafe
Overall judgment	Inspected All rank A	Caution One or more rank B	Unsafe One or more rank C

$$I_s = E_0 \times S_D \times T \tag{2.1}$$

where;

E_0 seismic capacity index and calculated with Eq. (2.2).

S_D unbalance index.

T Aging index

$$E_0 = \frac{1}{A_i} \times C \times F \tag{2.2}$$

where;

A_i Restoring force distribution shape factor in the vertical direction.

C Strength index.

F Ductility index.

Secondly, the seismic index with considering the damage, $_d I_s$, is calculated. The strength index, C, is reduced according to the seismic capacity reduction factor, η, which is defined according to the damage class, as shown in Table 2.4. The factor is defined based on the ratio of residual strain energy, E_r, to the total strain energy, $E_d + E_r$ in Fig. 2.2.

Table 2.4 Seismic capacity reduction factor, (η Bunno et al. 2006)

Damage class	Flexural member	Shear member
I	0.95	0.95
II	0.75	0.60
III	0.50	0.30
IV	0.10	0
V	0	0

The residual seismic capacity index, R, is defined as the ratio of the reduced seismic index with the reduction factors and the seismic index at the original condition (w/o damage), as shown in Eq. (2.3). According to the value of R, the damage level of the building is classified as "no damage" (R = 1.0), "Slightly damaged" (R > 0.95), "Minor damage" (0.95 > R > 0.80), "Moderate damage" (0.80 > R > 0.60), "Severe damage" (0.60 > R), and "Collapse". The decision of whether to demolish or repair the damaged building is made according to the matrix of both damage level and seismic intensity. For example, even if the damage level is "slight damage", the repairment is not recommended if the seismic intensity at the site is small (less than 5+ according to the standard).

$$R = \frac{dI_s}{I_s} \tag{2.3}$$

Since it takes at least several weeks and costs a lot to calculate I_s and $_dI_s$, a simplified function is also proposed in the standard. Each vertical member is grouped as a. Flexural member, b. Shear member, c. Plane wall, d. Plane wall with one boundary column, and e. Plane wall with boundary columns at both ends. The strength ratio among the groups is assumed as a:b:c:d = 1:1:2:6. The I_s and $_dI_s$ are calculated with the assumed strength C, seismic capacity reduction factor shown in Table 2.4, and the ductility factor F (=1).

2.4   Loss Estimation for Earthquake Insurance

The earthquake insurance in Japan was developed in 1966 after the 1964 Niigata Earthquake. The insurance aims are to compensate for the loss of houses and home contents to support rebuilding the daily life. The insurance is funded by the government. In this paper, the earthquake insurance for the loss of houses is introduced. When the insurance system was developed, the insurance payment was placed only for the totally collapsed houses. It was changed to have three categories, collapse, half-collapse, and partially collapse, and the amount of payment was decided according to the categories. After the 2011 Tohoku Earthquake, the category was changed to entirely damaged, partially damaged+, partially damaged-, and minor damage. The assigned inspector conducts the estimation.

The category is derived according to the sum of the damage ratio, as shown in Table 2.5. If the sum of the damage ratio is equal to or greater than 50%, 100% of the earthquake insurance premium is paid. The 60, 30, and 5% of the earthquake insurance premium are paid if the damage is categorized as partially damaged+, partially damaged-, and minor damage, respectively.

Firstly, the damage ratio of the whole building is calculated according to the settlement and inclination, as shown in Table 2.6. If the maximum settlement is greater than 100 mm, or if the inclination is greater than 2.1/100, the damage is categorized as "entirely damaged", without calculating damage ratio.

Table 2.5 Earthquake Insurance premium according to the damage (The general insurance association of Japan 2019)

	Damage class	Compensated damage	Insurance premiums paid
Building	Entirely damaged)	The loss percentage of the structure due to an earthquake becomes equal to or greater than 50% of the building	100% of Earthquake insurance premium (up to the actual value of the building)
	Partially damaged+	The loss percentage of the structure due to an earthquake becomes 40–50% of the building	60% of Earthquake insurance premium (up to 60% of the actual value of the building)
	Partially damaged−	The loss percentage of the structure due to an earthquake becomes 20– 40% of the building	30% of Earthquake insurance premium (up to 30% of the actual value of the building)
	Minor damage	The loss percentage of the structure due to an earthquake becomes 3–20% of the building	5% of Earthquake insurance premium (up to 5% of the actual value of the building)

Secondly, the damage ratio of the member damage is calculated. The ratios are listed in Table 2.7, according to the ratio of the number of each damage class to the total number. The ratio is calculated for the most damaged floor. The damage class of members is the same as the damage classification, but the damage classes IV and V are merged into one class of IV. If the number of the damage class IV members exceeds 50%, the building is categorized as "Entirely damaged", and the damage ratio is not calculated.

2.5   The Structural Health Monitoring System

2.5.1   Outline of the System

The outline of the evaluation is shown in Fig. 2.3 (Kusunoki 2016, 2018; Kusunoki and Teshigawara 2003, 2004; Kusunoki Et Al. 2008, 2012, 2018). The maximum responses during a main shock and aftershock are estimated as the intersection of the capacity and demand curves. The capacity curve is the relationship between the representative restoring force and representative displacement, which are derived from the measured accelerations instrumented into the building, as Fig. 2.4. The demand curve is the relationship between the response acceleration spectrum and response displacement spectrum, which are derived from the acceleration at the basement of the building. The amount of the damping coefficient needs to be assumed when the demand curve is derived. The damping coefficient for the elastic stage can be assumed as the viscous damping ratio of 5% as "Curve 1" shown in Fig. 2.1. When

Table 2.6 Standard table for damage ratios due to settlement and inclination (The general insurance association of Japan 2019)

Damage	Damage ratio (%)	Damage	Damage ratio(%)
Damage of the building	Maximum settlement	① Greater than 5 cm and less than or equal to 10 cm	3
		② Greater than 10 cm and less than or equal to 15 cm	5
		③ Greater than 15 cm and less than or equal to 20 cm	10
		④ Greater than20cm and less than or equal to 25 cm	15
		⑤ Greater than 25 cm and less than or equal to 30 cm	20
		⑥ Greater than 30 cm and less than or equal to 40 cm	25
		⑦ Greater than 40 cm and less than or equal to 50 cm	30
		⑧ Greater than 50 cm and less than or equal to 60 cm	35
		⑨ Greater than 60 cm and less than or equal to 80 cm	40
		⑩Greater than 80 cm and less than or equal to 100 cm	45
		⑪Greater than 100 cm	Entirely damaged
	Inclination	① Greater than 0.2/100 (about 0.1°), and less than or equal to 0.3/100 (about 0.2°)	3
		② Greater than 0.3/100 (about 0.2°), and less than or equal to 0.6/100 (about 0.4°)	5
		③ Greater than 0.6/100(about 0.4°), and less than or equal to 0.9/100(about 0.6°)	10
		④ Greater than 0.9/100(about 0.6°), and less than or equal to 1.2/100(about 0.7°)	15
		⑤ Greater than 1.2/100(about 0.7°), and less than or equal to 1.5/100(about 0.9°)	20

(continued)

Table 2.6 (continued)

Damage	Damage ratio (%)	Damage	Damage ratio(%)
		⑥ Greater than 1.5/100(about 0.9°), and less than or equal to 1.8/100(about 1.1°)	30
		⑦ Greater than 1.8/100(about 1.1°), 2.1/100(about 1.2°)	40
		⑧ Greater than 2.1/100(about 1.2°)	Entirely damaged

the building experience yielding as point (A) in Fig. 2.3, an additional damping effect due to non-linear response needs to be considered. Since the additional damping effect increases corresponding to the damage of the building, the total damping coefficient increases according to the representative displacement. Therefore, the demand curve is reduced from point (B) as "Curve 2" in Fig. 2.3. The maximum response during the main shock is predicted as the intersection of the capacity curve and the reduced demand curve (Curve 2), point (C) in Fig. 2.3.

On the other hand, the same method can be applied to predict the maximum response during an aftershock with considering the main shock and the following aftershock as one very long duration earthquake. The input energy of the combined earthquake is consequently larger than that of the main shock; then the maximum response may be larger than that of the main shock. It means that the equivalent damping effect becomes smaller than that of only the main shock as "Curve 3" shown in Fig. 2.3. The predicted maximum response during the aftershock is the intersection of Curve 3 and the capacity curve, with the assumption that the maximum aftershock is the same as the main shock.

In order to evaluate the safety of the building, the first mode of the response needs to be taken out to derive the capacity curve. The ultimate point is defined with the safety limit of each story. The maximum story drift of each story is derived from the maximum representative displacement and the first mode shape. Since the proposed safety evaluation is based on the first mode, the higher mode effect needs to be considered separately, if the higher mode effect is negligible, such as high-rise buildings (Fig. 2.5).

2.5.2 Capacity Curve from the Measured Acceleration

The representative acceleration $_s\ddot{\Delta} + \ddot{x}_0$ and representative displacement $_s\Delta$ can be derived using Eqs. (2.3) and (2.4) based on the first mode, respectively (Kusunoki et al. 2012):

Table 2.7 Standard table for damage ratios due to member damage (The general insurance association of Japan 2019)

Damage	Damage condition (Physical damage ratio)	Damage ratio (%)	
I	Fine cracks that can be seen in close distance	① Less than or equal to 10%	0.5
		② Greater than 10% and less than or equal to 20%	1
		③ Greater than 20% and less than or equal to 30%	2
		② Greater than 30% and less than or equal to 40%	3
		② Greater than 40% and less than or equal to 50%	4
		② Greater than 50%	5
II	Cracks are clearly visible	① Less than or equal to 5%	0.5
		② Greater than 5% and less than or equal to 10%	1
		③ Greater than 10% and less than or equal to 15%	2
		④ Greater than 15% and less than or equal to 20%	4
		⑤ Greater than 20% and less than or equal to 25%	5
		⑥ Greater than 25% and less than or equal to 30%	6
		⑦ Greater than 30% and less than or equal to 35%	8
		⑧ Greater than 35% and less than or equal to 40%	9
		⑨ Greater than 40% and less than or equal to 45%	10
		⑩ Greater than 45% and less than or equal to 50%	11
		⑪ Greater than 50%	13
III	Concrete partially crushes, there are wide cracks, and rebar/steel can be seen	① Greater than 3%	2
		② Greater than 3% and less than or equal to 5%	3
		③ Greater than 5% and less than or equal to 10%	5
		④ Greater than 10% and less than or equal to 15%	8
		⑤ Greater than 15% and less than or equal to 20%	10

(continued)

Table 2.7 (continued)

Damage	Damage condition (Physical damage ratio)	Damage ratio (%)	
		⑥ Greater than 20% and less than or equal to 25%	13
		⑦ Greater than 25% and less than or equal to 30%	15
		⑧ Greater than 30% and less than or equal to 35%	18
		⑨ Greater than 35% and less than or equal to 40%	20
		⑩Greater than 40% and less than or equal to 45%	23
		⑪Greater than 45% and less than or equal to 50%	25
		⑫Greater than 50%	30
IV	There are many wide cracks, cover concrete falls down a lot, and core concrete gets damaged, and reinforcement is visible Rebar buckled, and even core concrete falls down	① Less than or equal to 3%	3
		② Greater than 3% and less than or equal to 5%	5
		③ Greater than 5% and less than or equal to 10%	9
		④ Greater than 10% and less than or equal to 15%	14
		⑤ Greater than 15% and less than or equal to 20%	18
		⑥ Greater than 20% and less than or equal to 25%	23
		⑦ Greater than 25% and less than or equal to 30%	27
		⑧ Greater than 30% and less than or equal to 35%	32
		⑨ Greater than 35% and less than or equal to 40%	36
		⑩Greater than 40% and less than or equal to 45%	41
		⑪Greater than 45% and less than or equal to 50%	45
		⑫Greater than 50%	Entirely damaged

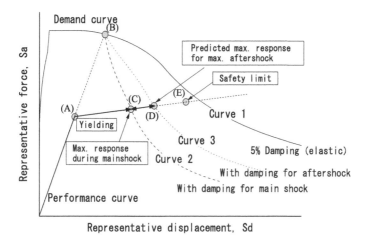

Fig. 2.3 Performance curve and demand curve (Kusunoki et al. 2018)

Fig. 2.4 Configuration of
the monitoring

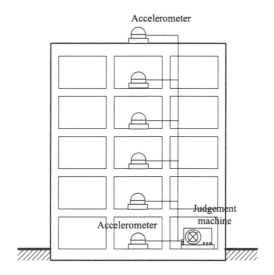

$$\left(_1\ddot{\Delta} + {}_1\ddot{x}_0\right) = \frac{\sum m_i \cdot {}_1x_i{}^2}{\left(\sum m_i \cdot {}_1x_i\right)^2} \cdot \sum_{i=1}^{N} {}_1P_i \tag{2.3}$$

$$_1\Delta = \frac{\sum m_i \cdot {}_1x_i{}^2}{\sum m_i \cdot {}_1x_i}. \tag{2.4}$$

The representative displacement can be obtained from Eq. (2.4) by using the relative displacement obtained from the predominant displacement time histories.

The shape of the external force distribution $_1P$ of Eq. (2.3) should be proportional to the first mode vector. In order for the absolute acceleration to be proportional to the

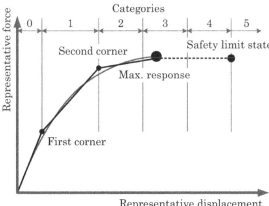

Fig. 2.5 Capacity curve and category of its maximum representative displacement (Kusunoki et al. 2018)

first mode vector, the stimulation factor $_1\beta \cdot \{_1 u\}$ of the first mode must be multiplied by the ground acceleration $_1\ddot{x}_0$. This means that the first mode of the unit vector $\{1\}$ is multiplied by the ground acceleration. As a result, the external force proportional to the first mode vector is obtained as.

$$_1 P_i = m_i(_1\ddot{x}_i + _1\beta \cdot _1 u_i \cdot _1\ddot{x}_0). \tag{2.5}$$

The representative acceleration in Eq. (2.6) is obtained by substituting Eq. (2.5) into Eq. (2.3):

$$\left(_1\ddot{\Delta} + _1\ddot{x}_0\right) = \frac{\sum m_i \cdot _1 x_i^2}{\left(\sum m_i \cdot _1 x_i\right)^2} \sum_{i=1}^{N} m_i \cdot _1\ddot{x}_i + _1\ddot{x}_0. \tag{2.6}$$

As shown in Eq. (2.6), only the relative acceleration term of the representative acceleration is required to be divided by the equivalent mass ratio when the representative acceleration is derived from the measured accelerations.

In Eqs. (2.4) and (2.6), the order of the mass m_i is the same in the denominator and the numerator. Therefore, we require the mass ratio between floors instead of the absolute mass. If the usage of the building is the same for all floors, the floor-area ratio can be used instead of the mass ratio.

2.6 Target Building

The proposed health monitoring system is installed into the building for the department of architecture of Yokohama National University at the beginning of the year of 2008. The building has eight stories and one underground floor. The height of the building is 30.8 m, and its structural type is steel-reinforced concrete. The building was designed before 1981 when the Japanese building code was revised to confirm the ultimate strength of buildings. It was found that the building did not have enough ultimate strength, and then the building was retrofitted. The retrofitting construction had been conducted from July 2008 to May 2009, and the sensors were removed at that time. The building before and after retrofitting is shown in Fig. 2.6. The key plan is shown in Fig. 2.7. EW direction is the longitudinal direction, and NS direction is transverse direction.

(a) Before retrofitting   (b) After retrofitting

Fig. 2.6 Instrumented building

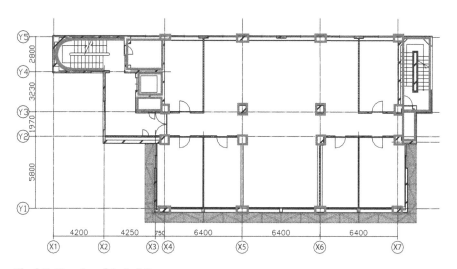

Fig. 2.7 Key plan of the building

After starting the monitoring, 112 earthquakes responses are measured until 2011 Off the Pacific Coast of Tohoku Earthquake, which occurred at 14:36, March 11th, 2011. After that, about 530 earthquake records are measured until the end of 2011.

2.7   Response During the 2011 Tohoku Earthquake

The health monitoring system worked well during 2011 Off the Pacific Coast of Tohoku Earthquake (Kusunoki Et Al. 2018). Figure 2.8. shows the measured lateral accelerations on the basement and roof. The maximum acceleration was 91.5 cm/s^2 on the basement and 410 cm/s^2 on the roof. The predominant component of the acceleration lasted about 180 s.

The measured performance curve, skeleton curve from the performance curve, and the demand curve in the EW direction are shown in Fig. 2.9. The vertical axis of the demand curve is the response absolute acceleration spectrum Sa, and the horizontal axis is the response displacement spectrum Sd with the viscous damping

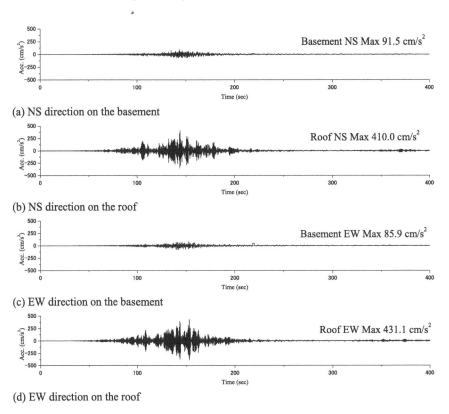

(a) NS direction on the basement

(b) NS direction on the roof

(c) EW direction on the basement

(d) EW direction on the roof

Fig. 2.8 Measured earthquake during 2011 Off the Pacific Coast of Tohoku Earthquake

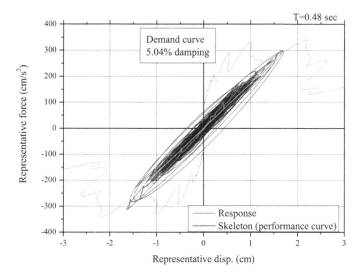

Fig. 2.9 Measured performance and demand curves during 2011 Off the Pacific Coast of Tohoku Earthquake (EW direction)

factor of 5%. The maximum representative displacement of 1.7 cm was measured in the positive direction. The equivalent period from the maximum displacement point in the positive direction was 0.48 s. The calculated viscous damping for the demand curve in order to get the same demand value for the period of 0.48 as the maximum response was 5.04%, which is a reasonable value.

Since the natural period in the EW direction before the earthquake was about 0.41 s, the equivalent period of 0.48 is longer than the period before the earthquake. Figure 2.10 shows the skeleton curve and the slopes for the periods of 0.41 and 0.48 s. It is clearly found that the stiffness degrading started at the representative acceleration of about 100 cm/s². The stiffness degraded down to 73% according to the change of the period from 0.41 to 0.48 s.

From Fig. 2.10, it can be said that the frequency change can be observed more accurately from the performance curve than from the transfer function since the slope of the performance curve is square of the predominant angular frequency ω. The transfer function sometimes does not show any predominant frequency if a large nonlinearity occurs during an earthquake. Moreover, while the performance curve shows the building has not yielded yet, it is unclear whether the damage is serious only from the frequency change.

After the main shock, cracks occurred in the building were investigated. The observed cracks in the Y3 frame are shown in Fig. 2.11. Cracks occurred mainly at the bottom of the continuous shear walls and at the corner of openings. These cracks probably cause the stiffness degradation of the performance curve shown in Fig. 2.11.

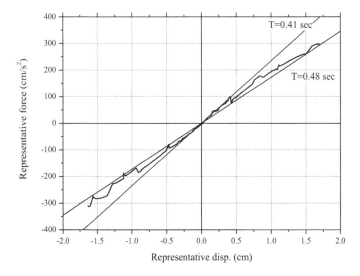

Fig. 2.10 Skeleton curve of the measured performance during 2011 Off the Pacific Coast of Tohoku Earthquake (EW direction)

Fig. 2.11 Observed cracks in the Y3 frame (EW direction)

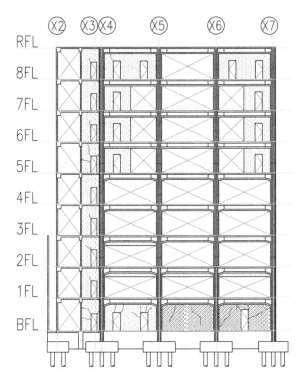

2.8   Conclusions

The rapid inspection method, the damage classification method, and the loss classification method for earthquake insurance, which are all based on the visual inspection and applied in Japan, are introduced in this chapter. Recent earthquakes revealed that visual inspection is hard to conduct because most of all structural members are covered by finishing, especially for high-rise buildings. Right after an earthquake, it is quite difficult to grasp the outline of the damage, which is needed to decide the target area to inspect. Sensing technology probably helps a lot to overcome the problems. The Ministry of Land, Infrastructure, and Transportation of Japan organized a committee to discuss how to apply the structural health monitoring system for the rapid inspection. The general insurance association of Japan organized a committee as well to discuss how to apply it for shortening the duration to decide the amount of the insurance payment. The sensing technology will be applied widely in the field of disaster reduction soon. Research to bridge the structural health monitoring result and existing inspection method will be needed.

References

Bunno M, Maeda M, Nagata M (2006) Damage classification method for the damaged R/C buildings based on the residual seismic capacity of structural members. Proc Japan Concr Inst 22(3):1447–1452 (in Japanese)

Japan building disaster prevention association (2015) Guideline for post-earthquake damage evaluation and rehabilitation of RC buildings in Japan Part II RC and SRC buildings (in Japanese)

Kusunoki K (2016) Damage evaluation of a base-isolated building with measured accelerations during Tohoku Earthquake. In: The 16th world conference on earthquake engineering, digital

Kusunoki K (2018) A new structural health monitoring system for real-time evaluation of building damage. Seism Haz Risk Assess 331–343. https://doi.org/10.1007/978-3-319-74724-8_22

Kusunoki K, Teshigawara M (2003) A new acceleration integration method to develop a real-time residual seismic capacity evaluation system. J Struct Constr Eng 569:119–126 (in Japanese)

Kusunoki K, Teshigawara M (2004) Development of real-time residual seismic capacity evaluation system—integral method and shaking table test with plain steel frame. In: The 13th world conference on earthquake engineering, CD-Rom

Kusunoki K, Elgamal A, Teshigawara M, Conte JP (2008) Evaluation of structural condition using Wavelet transforms. In: The 14th world conference on earthquake engineering, CD-Rom

Kusunoki K, Tasai A, Teshigawara M (2012) Development of building monitoring system to evaluate residual seismic capacity after an earthquake. In: The 15th world conference on earthquake engineering, digital

Kusunoki K, Hinata D, Hattori Y, Tasai A (2018) A new method for evaluating the real-time residual seismic capacity of existing structures using accelerometers: structures with multiple degrees of freedom. Japan Architect Rev, Architectural Institute of Japan

The general insurance association of Japan (2019) Contract bookmark for earthquake insurance (revised in January 2019) (in Japanese)

Chapter 3
Post-earthquake Demolition in Christchurch, New Zealand: A Case-Study Towards Incorporating Environmental Impacts in Demolition Decisions

R. E. Gonzalez, M. T. Stephens, C. Toma, K. J. Elwood, and D. Dowdell

Abstract The 2010/2011 Canterbury Earthquake Sequence resulted in severe loss and disruption in Christchurch, New Zealand due to liquefaction and damage from strong shaking. Following the earthquake, over 60% of concrete buildings with 3 + stories in the Christchurch CBD were demolished, resulting in a widespread displacement of people and business, an excess of \$NZD 40 billion in losses, and significant environmental impacts from the demolition. Following the event, it was revealed that environmental impacts were not a direct consideration in demolition decision making. This paper provides a quantitative evaluation of the environmental impacts of the demolitions in Christchurch to highlight the importance of including environmental considerations when deciding between repair or demolition of a damaged building. First, the quantitative and qualitative factors that led to the demolitions following the Canterbury Earthquake Sequence are discussed to provide context for the argument that environmental impacts should be included in such considerations. Next, the environmental impacts of building demolitions in Christchurch are presented in terms of the embodied CO_2 and energy in the building materials; the demolition process and waste disposal are not considered in this initial evaluation. Finally, a brief discussion on incorporating environmental impacts into the demolition decision making paradigm is presented. Moreover, consideration of environmental impacts of demolitions supports the need to move toward low-damage design in the future evolution of building codes.

R. E. Gonzalez (✉) · M. T. Stephens · C. Toma · K. J. Elwood
University of Auckland, Auckland 92019, New Zealand
e-mail: rgon868@aucklanduni.ac.nz

D. Dowdell
Building Research Association of New Zealand, Porirua 5240, New Zealand

© The Author(s) 2021

S. Akkar et al. (eds.), *Advances in Assessment and Modeling of Earthquake Loss*,
Springer Tracts in Civil Engineering,
https://doi.org/10.1007/978-3-030-68813-4_3

3.1 Introduction

The 2010/2011 Canterbury Earthquake Sequence caused severe loss and disruption
in Christchurch, New Zealand. The most damaging event (M6.2, 5-km deep, 10-km
SE of the Christchurch Central Business District (CBD)) occurred 22 February 2011
and resulted in 185 fatalities and widespread impacts on the community including
loss of land and infrastructure due to liquefaction and damage from strong shaking
and a cordon (public exclusion zone) that restricted access to a majority of the CBD
which resulted in a widespread displacement of people and businesses (Potter et al.
2015). In the fallout from the earthquake, over 60% of concrete buildings with 3 +
stories in the Christchurch CBD were demolished, resulting in an excess of $NZD
40 billion in losses (Parker and Steenkamp 2012). The high rate of demolition,
particularly among buildings with relatively little damage, highlighted the complex
quantitative (e.g. building damage, year of construction, occupancy) and qualitative
factors (e.g. insurance and legislation) that influence the repair-or-demolish decision
faced by building owners following an event (Kim et al. 2017; Marquis et al. 2017). In
addition, Marquis et al. (2017) highlighted that the environmental impacts associated
with demolishing a damaged building prior to the end of its design life—in terms of
embodied CO_2 and energy, waste generation during demolition and greenhouse gas
emissions resulting from the construction of a new building—were largely ignored
in the decision-making process, which is contrary to the principles of sustainability
advocated by stakeholders, policymakers, and the general public (Carbon Leadership
Forum 2019; MBIE 2020).

To begin to highlight the importance of incorporating environmental considerations in demolition decision making following earthquakes, this paper presents
a preliminary study that provides a quantitative evaluation of the environmental
impacts of building demolitions in Christchurch following the Canterbury earthquakes. First, the quantitative and qualitative factors that led to the demolitions are
discussed to provide context for the argument that environmental impacts should
be included in such considerations. Next, the environmental impacts of building
demolitions in Christchurch are presented in terms of embodied CO_2 and energy
in the building materials. Finally, a brief discussion on incorporating environmental
impacts into the demolition decision making paradigm is presented. Note that this
paper is intended as a preliminary investigation into the environmental impacts
associated with the demolition of buildings following earthquakes. There are many
complex factors that contribute to the total environmental impact of the decision to
demolish or repair a building including: building materials, building age, construction processes, repair strategies (if repaired), demolition processes (if demolished),
and waste disposal. Only the embodied CO_2 and energy in the building materials are
considered here.

3.2   Factors that Influenced Demolition Decisions in Christchurch

The Christchurch CBD includes approximately 110 city blocks bounded by four avenues: Deans, Bealey, Fitzgerald, and Moorhouse as illustrated in Fig. 3.1. At the time of the earthquakes, there were approximately 3000 buildings within the CBD, consisting primarily of commercial, light industrial, and residential buildings (Kam and Pampanin 2011). The dominant multi-story typologies of commercial and multi-unit residential buildings within the CBD were reinforced concrete moment frame and shear wall buildings designed to "modern" (mid-1980's and onwards) provisions for seismic resistance. From the perspective of design performance and life safety, these dominant typologies tended to perform satisfactorily during the earthquakes, with plastic hinges forming in discrete regions, allowing the buildings to dissipate energy and people to evacuate. Further, a large number of buildings within these typologies had relatively little damage following the event (refer to Fig. 3.3) (Kim et al. 2017). However, in September 2014, the Canterbury Earthquake Recovery Authority (CERA) reported that 150 "significant" buildings (generally commercial and multi-unit residential 5 + storeys) had been demolished in the Christchurch CBD, representing approximately 65% of significant buildings in the CBD and immediately surrounding neighbourhoods (Marquis et al. 2017). The geographical distribution of building demolitions within the Christchurch CBD is shown in Fig. 3.1.

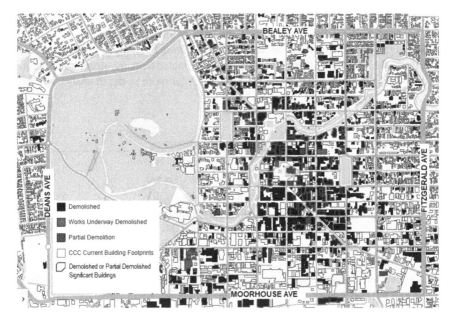

Fig. 3.1 Overview of building demolitions in christchurch CBD—November 2014 (figure reproduced from (Marquis et al. 2017))

The apparent disconnect between the level of damage and the high number of demolitions in Christchurch following the Canterbury Earthquake Sequence highlighted a lack of understanding surrounding the decision making process that leads to the demolition or repair of a structure—how was it that so many relatively lightly damaged buildings were demolished? To answer this question, several studies were initiated to evaluate the complex quantitative and qualitative parameters that factor into the decision to demolish a damaged building following an event (Kim et al. 2017; Marquis et al. 2017). These parameters are discussed in the following sections.

3.2.1 Quantitative Factors

Prior to the Canterbury Earthquake Sequence, there were no comprehensive quantitative studies on building demolitions following earthquakes. Kim et al. (2017) used a building data set consisting of 223 RC moment frame and shear wall buildings (herein referred to as *Building Set A*) in Christchurch to identify quantitative variables that influenced post-earthquake decisions on whether to demolish or repair the buildings; these buildings represented approximately 88% of RC buildings of 3 storeys and higher in the Christchurch CBD. The variables for each building included building identification information, decision outcome, damage indicators, building condition, lateral load resisting system (LLRS), duration in cordon, year of construction, heritage status, footprint area, number of floors, and occupancy. A brief overview of the key parameters (Table 3.1) are discussed here, with more detailed descriptions of all parameters in reference material (Kim 2015; Kim et al. 2017).

The overall building demolition statistics for *Building Set A* are shown in Fig. 3.2, while the building demolition statistics based on estimated damage, structural system, and duration in the cordon are shown in Fig. 3.3. A far more in-depth analysis of the

Table 3.1 Building parameters

Parameter	Measure/description
Decision outcome	Demolish, Repair, Unknown
Damage indicators	Damage Ratio[1], Placard[2]
Building condition	%NBS[3], Design ductility[4]
Lateral Load Resisting System (LLRS)	Moment Frame (MF), Shear Wall (SW), Moment Frame and Infill (MFIF), Combined MF and SW
Duration in cordon	Number of months in cordon

[1]Approximate visual estimate of building damage as a ratio of repair cost to replacement cost
[2]Usability of assessed building (Green—inspected, Yellow—Restricted use, Red—Unsafe)
[3]Lateral load carrying capacity of a building as a function of the current building standard. Only available for 15 buildings
[4]Only available for 15 buildings

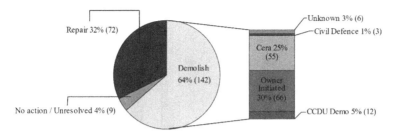

Fig. 3.2 **a** Overall demolition statistics for buildings in buildings set and **b** Party responsible for demolition

demolition data for the full range of parameters presented in Table 3.1 can be found in Kim (2015) and Kim et al. (2017).

A total of 64% of the buildings in the building set were demolished (142 buildings representing 64% of the total floor space), 32% were repaired (72 buildings representing 34% of the total floor space), and the fate of 4% were unknown at the time of data collection. As a fraction of the total number of buildings in the building set, 2% (3 buildings) were demolished immediately following the earthquake under instruction from *Civil Defence* for public safety, 5% (12 buildings) were demolished to make way for other revitalisation projects, 25% (55 buildings) were demolished under instruction from CERA, and 30% (66 buildings) were demolished based on the decision of the owner.

Figure 3.3a and b illustrates the building demolition statistics for varying levels of building damage as estimated using rapid assessments following the earthquake. The damage metrics evaluated included placard posting (green, yellow, red) and damage ratio (approximate visual estimate of building damage as a ratio of repair cost to replacement cost, excluding contents). Within the building set, 35% (77 buildings) received green placards, 46% (103 buildings) received yellow placards, while 19% (43 buildings) received red placards. In addition, among 61% (135 buildings) to have been assessed a relatively low damage ratio of 10% or less, 47% (63 buildings) were demolished. These statistics would appear to indicate that a significant number of buildings with relatively low damage were demolished.

To provide a slightly deeper look at the data here, Fig. 3.3c and d further segregate the data to evaluate the relationship between lateral load resisting system, damage ratio, duration in the cordon, and demolition statistics. RC moment frame and shear wall lateral load resisting systems were equally represented in the building data set (40% and 44% respectively), however moment frame buildings had a much higher rate of demolitions (75%) than shear wall buildings (49%). A high rate of demolition (75%) was observed for buildings that remained in the cordon for more than 6 months, and an additional study (Chang et al. 2014) noted that being located within the cordon facilitated ease of demolition and may have led to a higher percentage of buildings in the cordon being demolished than would have been warranted based on damage alone.

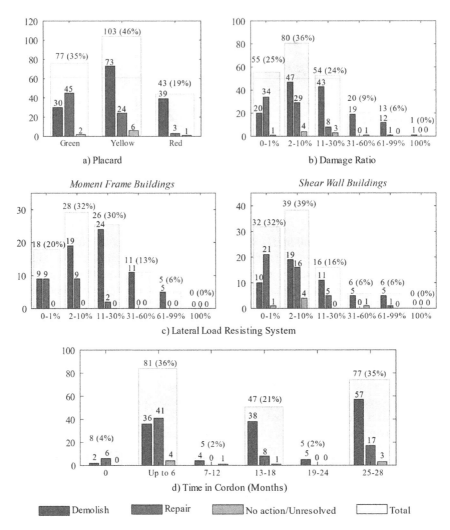

Fig. 3.3 **a** Placard placement demolition statistics, **b** Damage ratio demolition statistics, **c** Lateral load resisting system demolition statistics, and **d** Duration in cordon damage statistics for buildings in the building set

3.2.2 Qualitative Factors

The high demolition rate of relatively undamaged buildings in Christchurch (as summarised in the previous section) revealed complex and unique issues facing owners of commercial and multi-storey residential buildings when deciding the future of earthquake-impacted buildings. In particular, it became clear that while damage level is a good indicator of the seismic performance of a building, other multifaceted variables are involved in the ultimate decision to repair or demolish.

Marquis et al. (2017) used a reduced building set of 15 buildings (herein referred to *Building Set B*) to explore the factors beyond structural damage that led to the decision to demolish or repair. The buildings in *Building Set B* were distributed throughout the Christchurch CBD and surrounding neighborhoods and were representative of the entire RC building stock within the CBD. Approximately half the buildings in *Building Set B* were repaired (7 buildings), while the remainder were demolished (8 buildings). Through this more detailed analysis, several qualitative factors that influence whether a building is demolished or repaired were revealed. Three of these factors are summarised here: (1) insurance, (2) building legislation, and (3) owner preference. Further details are provided in Marquis (2015) and Marquis et al. (2017).

3.2.2.1   Insurance

Approximately 80% of the economic loss from the Canterbury Earthquakes was covered by the insurance industry, which is considerably higher than other recent earthquakes worldwide (as illustrated in Fig. 3.4) (Bevere and Grollimund 2012). This high level of insurance penetration and unique policy wording played a critical role in the high rate of demolitions of relatively undamaged buildings in Christchurch (Kim et al. 2017; Marquis et al. 2017). The majority of commercial buildings in Christchurch were insured under a reinstatement policy, which entitles the policy owner to receive a repaired property which is largely the same in appearance, quality, and working order as "when new". However, following the event, it was revealed that the sum insured was frequently less than the actual rebuilding cost for most commercial properties, which meant the policy was not adequate to provide replacement of the building. The inadequate coverage of many commercial buildings in the CBD rendered technically viable repair and/or strengthening work uneconomic and resulted in a large number of cash settlement payouts and subsequent demolitions. All of the buildings in *Building Set B* cash settled. It should be noted that the unusually high rate of insurance penetration, combined with low deductibles and specific requirements in the reinstatement policies (e.g. the "as when new" clause), made insurance a relatively New Zealand specific factor in building demolition decisions following the Christchurch Earthquake (Marquis et al. 2017).

Fig. 3.4 Economic losses covered by insurance in recent earthquakes

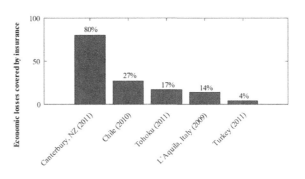

3.2.2.2 Building Legislation

The Building Act (2004) requires all new building work in New Zealand to comply with the New Zealand Building Code (Department of Building and Housing 2011). The Act applies to the construction of new buildings as well as the alteration and demolition of existing buildings. The Act deems a building to be *earthquake prone* if it has a lateral capacity ≤33% of a building designed and constructed to the current building code or New Building Standard (NBS). It should be noted that the Building Act does not explicitly consider the repair of earthquake-damaged buildings. Instead, earthquake damage is considered as an alteration, and as a result, the assessment and upgrade of fire systems and accessibility features are also triggered when earthquake damage is repaired, which can increase the effective repair costs substantially.

There were two changes to the building regulations following the earthquakes that influenced demolition decisions on earthquake prone buildings. First, the Christchurch City Council attempted to change the earthquake-prone building policy to require that building owners strengthen their buildings to 67% NBS, as opposed to 34% pre event. This change was struck down by a High Court decision in 2013 (and a Supreme Court decision released in December 2014), which stated that owners are only required to strengthen buildings to 34% NBS. However this reversal resulted in confusion as to (1) whether insurers were required to pay for the additional remediation (New Zealand Supreme Court 2014) and (2) whether building owners were required to upgrade to 67% NBS to receive a building consent for earthquake repairs. Second, the seismic hazard factor in Canterbury was increased from 0.22 to 0.3 to take into account heightened seismicity in the region following the earthquakes. This effectively forced down the %NBS rating of many buildings in Christchurch. Both these changes have had a significant effect on post-earthquake decisions and the cost of the repair (and strengthening), which may have led to more building demolitions than would have occurred without the legislation changes.

The influence of %NBS on demolition decisions is illustrated using *Building Set B* (Fig. 3.5). All buildings that were repaired in *Building Set B,* except one, had %NBS values above 67%, while all demolished buildings had %NBS below 66%, including two with %NBS below 33% (which would be designated as *earthquake prone* in New Zealand).

3.2.2.3 Owner Preference

Although insurance and building legislation certainly played a role in owners' decisions to demolish buildings following the earthquakes, it is important to highlight owner preference as a stand-alone factor that led to demolitions due to the relatively high rate of owner-initiated demolitions (as illustrated in Fig. 3.2). In general, owner preference with regard to demolition was informed by both pre-earthquake economic conditions as well as unique decision-making strategies to avoid economic loss post-event (Marquis et al. 2017). Prior to the event, there was a significant surplus of commercial space in the Christchurch CBD, which contributed to a low

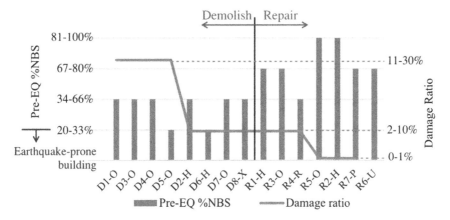

Fig. 3.5 %NBS and damage ratio for reduced building set of 15 buildings (code on x-axis refers to building ID from Marquis (2015) and Marquis et al. (2017)) (figure reproduced from Kim et al. (2017))

rent commercial office market relative to Wellington and Auckland. Thus, the lower income streams generated from office buildings may have incentivized owners to demolish and rebuild differently or invest elsewhere, rather than repair following the earthquake.

Following the event, owners were faced with a number of scenarios regarding the future of their building: the building could be repaired to the same performance level, the building could be repaired to a higher standard, the building could be demolished and replaced with an equivalent building, or the building could be demolished and not replaced. In general, it was found that owner decisions were based on economics with several exceptions. Additional owner decision-making variables including business strategy, perception of risk, uncertainty, building regulations (e.g. changes in the building code, compliance issues, etc.), and government decisions (e.g. cordons, mandatory demolitions, etc.). Through interviews, it was revealed that most building owners considered it a good outcome if their building was declared a total loss and demolished, because of the financial benefits, flexibility, and speed of cash settlements. Note, however that different decision strategies were observed for some heritage building owners, where there was a preference to refurbish an old building to preserve unique architectural features or emotional attachments.

3.2.3 Conceptual Demolish/Repair Framework

Based on the quantitative and qualitative factors that contributed to post-earthquake decision making in Christchurch, Marquis et al. (2017) developed a comprehensive multi-phase framework (illustrated in Fig. 3.6) that summarises the demolish/repair

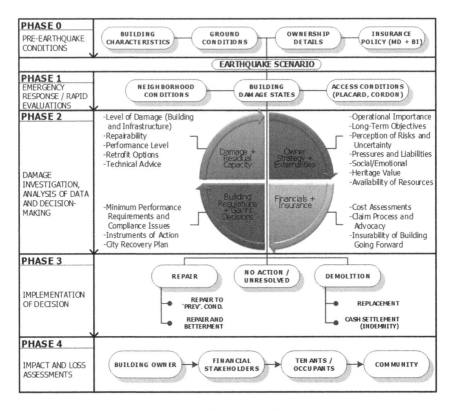

Fig. 3.6 Conceptual framework that identifies variables influencing post-earthquake decisions (figure reproduced from marquis et al. (2017))

decision making processes during and following the Canterbury earthquakes. Significant themes in the framework that were observed to be key variables that influenced decision making include: insurance, damage and residual capacity, decision making strategies, and legislation. The framework utilises a holistic approach by including a contextualisation phase (phase 0) that accounts for unique factors in the built environment or socio-economic factors that may have influenced the decision to repair or demolish. The additional phases in the framework represent the time immediately following the event through the implementation of the repair or demolition of the building.

Notably missing from the conceptual framework in Fig. 3.6 is any consideration of environmental impacts arising from building demolitions. Such impacts could be indicated in Fig. 3.6 under "Externalities" but were not highlighted by any stakeholders interviewed by Marquis et al. (2017) as influencing decision making. It is more likely that "Government Regulations" designed to de-incentivize demolitions based on environmental impacts would be needed to force this consideration into the decision-making framework. To rationalize any such regulations, it is critical to demonstrate the expected environmental impacts from building demolitions as described in the next section for buildings in Christchurch.

3.3 Quantification of Environmental Impacts of Demolitions

The afore-summarised studies that evaluated the factors that contributed to the repair or demolition of buildings following the Canterbury earthquakes revealed that the environmental impacts of demolishing repairable buildings before the end of their design life were ignored in the decision-making process. To begin to highlight the importance of incorporating environmental considerations in the decision to demolish or repair a building following an event, this section provides a preliminary evaluation of the environmental impacts of building demolitions following the Canterbury Earthquakes. Figure 3.7 summarises a comprehensive framework for calculating the environmental impacts associated with demolishing a building. The environmental impacts in the framework are broken into three distinct modules, namely: (1) embodied CO_2 and energy in the building materials, (2) impacts of the processes used in construction of the building, and (3) impacts of the transport and waste management processes after demolition. This initial study focuses on the embodied CO_2 and energy in the building materials (module 1 in Fig. 3.7) assuming a like-for-like replacement of materials that were demolished considering only the manufacturing of the materials, whilst excluding wider impacts of construction methods, repair (as an alternative to demolition), and waste management (in the case of demolition). These more complex factors will be incorporated in future studies.

The 142 demolished buildings from *Building Set A* (described above) and BRANZ CO_2NSTRUCT v1.0. (BRANZ 2019) were used to perform the preliminary environmental impact study. BRANZ CO_2NSTRUCT provides estimates for embodied CO_2 and energy in common structural (e.g. in-situ concrete, precast concrete, structural steel, etc.) and nonstructural (e.g. insulation, walls, glass, paint, etc.) components

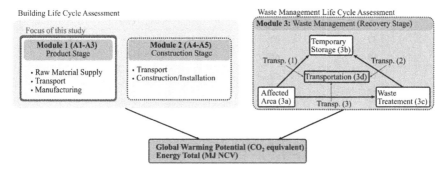

Fig. 3.7 Framework for calculating environmental impacts of building demolition

as a function of the total weight of the components in the building. The embodied carbon is a measure of greenhouse emissions converted to units of CO_2 equivalents which results from combustion of fossil fuels or chemical reactions that occur during material production, while the embodied energy considers fossil fuel, nuclear and/or renewable energy consumed in the material production (FEMA 2012).

To use the tool to determine the total embodied CO_2 and energy within the structural components of a concrete building, a material takeoff is required including the volume, material strength, and reinforcing ratio of in-situ and precast concrete components and shear walls. As detailed material takeoffs were not available for all 142 demolished buildings within *Buildings Set A*, a subset of 9 buildings were used to develop a material takeoff prediction model that could be applied across the entire building set. A summary of several key properties of the 9 buildings in the reduced set are given in Table 3.2 (including LLRS, number of storeys, year of construction and gross floor area (GFA)), while material takeoffs from the structural components in these buildings are summarised in Table 3.3 and Fig. 3.8. Note that several simpli-

Table 3.2 Reduced building set used to develop material takeoff model

Building	LLRS	Storeys	Year constructed	GFA, m^2
1	MRF	5	1986	1770
2	MRF	5	1976	1158
3	MRF	3	Pre 1965	895
4	MRF	12	1970	7914
5	SW	5	1976	1433
6	SW	4	1976	2703
7	SW	6	1978	1603
8	SW	9	1987	5151
9	MFIF	3	1975	617

Table 3.3 Material takeoff of reduced building set in percentage of total weight

Building	Foundations (%)	Beams, columns, slabs (%)	Precast (%)	Walls (%)
1	37	41	9	13
2	19	60	6	14
3	16	77	0	7
4	28	63	5	3
5	25	65	9	0
6	35	56	6	2
7	25	59	7	8
8	45	39	12	4
9	16	74	0	10

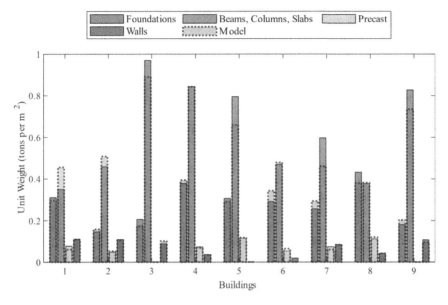

Fig. 3.8 Measured and estimated material takeoffs

fying assumptions were made when developing the material takeoffs. Namely, the volumetric reinforcing ratio, ρ_v, was assumed to be 1% for in-situ cast foundations, 1.5% for in-situ cast beams, columns and slabs, 1.5% for precast beams and columns and 1% for precast panels and slabs. In addition, concrete walls that were identified as part of the primary LLRS were designated as *reinforced* with a reinforcing ratio of 1.5% while infill concrete walls were designated as *unreinforced*. The strengths of all concrete components were assumed to be consistent for all components in each building, and were defined based on recommendations from (MBIE 2018) which provides estimates for concrete strength based on year of construction.

A linear regression was used to fit independent takeoff prediction models for each of the four material subsets described above (e.g. in-situ foundations with $\rho_v =$ 1%, in-situ beams, columns, slabs with $\rho_v = 1.5\%$, precast $\rho_v = 1.5\%$ (beams and columns), $\rho_v = 1\%$ (slabs and panels), and walls) as a function of gross floor area in the building. These regression models were combined into a single conditional regression model (Eq. 3.1), where *GFA* is the gross floor area in m^2, W_{mat} is the weight of the material in tonnes, and the coefficients A and B are a function of the characteristics of the building and are summarised in Table 3.4.

$$\text{Wmat} = \text{A} \times \text{GFA} + \text{B} \tag{3.1}$$

The accuracy of the material takeoff model was evaluated using the reduced set of 9 buildings taken from *Building Set A*. The predicted material takeoffs are compared to the measured takeoffs in Fig. 3.8. In general, the model was fairly effective in predicting the material takeoffs for the 9 buildings, with an average

Table 3.4 Regression coefficients for Eq. 3.1

Condition	A	B
Spread foundation (in-situ, $\rho_v = 1\%$)	0.108	58.91
Raft foundation (in-situ, $\rho_v = 1\%$)	0.328	56.67
Pile cap (in-situ, $\rho_v = 1\%$)	0.420	206.15
Beams and columns in building with structural walls (in-situ, $\rho_v = 1.5\%$)	0.275	551.83
Beams and columns in buildings without structural walls	0.744	– 305.52
Floor or panels (precast)	0.070	– 16.77
Beams or columns (precast)	0.053	0
Walls (with precast panels)	0.023	100.55
Walls (without precast panels)	0.117	– 13.03

percent difference of 4.5% across all materials. Larger discrepancies in individual materials (e.g. in building 1 for beam, columns, slabs) was a result of the use of precast elements not seen in the other buildings. Building 1 had precast slabs, which significantly reduced the measured volume of in-situ concrete for slabs, but increased the measured volume of precast concrete. The model respectively overpredicted and underpredicted these volumes.

The embodied CO_2 and energy in the structural components of the 142 demolished buildings from *Building Set A* were calculated using the material takeoff model in Eq. 1 in conjunction with BRANZ CO_2NSTRUCT. Due to limited information about the non-structural systems in the buildings, the contribution from non-structural components (including glazing and frames, mechanical, electrical and plumbing (MEP), and tenant improvements) were accounted for using values from literature (Carbon Leadership Forum 2019; Dowdell and Berg 2016). Note that data from the Carbon Leadership Forum (2019) was developed based on US building data.

The embodied CO_2 and energy calculated for each of the 142 demolished buildings are summarised Fig. 3.9, while contributions of the different material categories are summarised in Fig. 3.10. There was no apparent correlation between the primary LLRS and the embodied CO_2 or energy as indicated in Fig. 3.9, and the primary embodied CO_2 and energy contributor across the entire demolished building set was in-situ cast concrete as indicated in Fig. 3.10.

The total embodied CO_2 and energy across all 142 demolished buildings in *Building Set A* reveals the large environmental impact of post-earthquake demolitions. For context, the total embodied CO_2 across the demolished building set was 3.08×10^8 kg, which is approximately equal to the annual CO_2 emissions of purchased electricity in 400,000 average homes in New Zealand (Isaacs et al. 2010; Ministry for the Environment 2019). However, these values alone don't necessarily provide justification for incorporating environmental impacts into the decision to repair or demolish a building following an event, as some demolitions are unavoidable due to severe structural damage or residual deformation. To evaluate the environmental impacts of the demolition of relatively undamaged buildings in Christchurch,

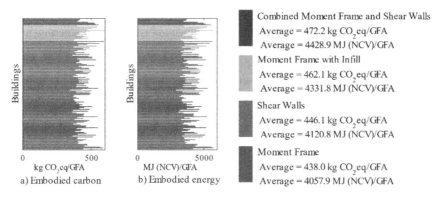

Combined Moment Frame and Shear Walls
Average = 472.2 kg CO_2eq/GFA
Average = 4428.9 MJ (NCV)/GFA
Moment Frame with Infill
Average = 462.1 kg CO_2eq/GFA
Average = 4331.8 MJ (NCV)/GFA
Shear Walls
Average = 446.1 kg CO_2eq/GFA
Average = 4120.8 MJ (NCV)/GFA
Moment Frame
Average = 438.0 kg CO_2eq/GFA
Average = 4057.9 MJ (NCV)/GFA

Fig. 3.9 **a** Embodied carbon and **b** Energy normalised by gross floor area of 142 demolished buildings from *Building Set A*

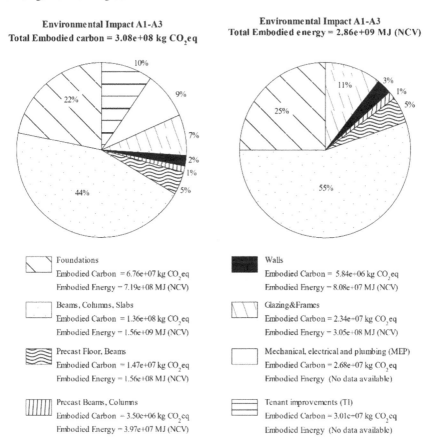

Fig. 3.10 Embodied carbon and energy across 142 buildings from building set A

the total embodied CO_2 of demolished buildings across *Building Set A* was evaluated as a function of the building damage ratios as shown in Fig. 3.11. A staggering 50% of the total embodied CO_2 in the Christchurch demolitions was from buildings with a damage ratio of less than 10%, which represents a very low damage threshold that could most likely have been repaired following the event. Had these buildings been repaired rather than demolished, the savings in the total embodied CO_2 across the demolished building set would have decreased significantly. However, the exact reduction cannot be calculated because there is no information regarding the type of repair which would have been necessary and the corresponding environmental cost. Considering the growing evidence that more heavily damaged buildings could likely have been repaired without safety concerns (Marder et al. 2020), it is not unreasonable to assume buildings with damage ratios up to 30% were repairable. Based on Fig. 3.11, this would suggest that nearly 80% of the total embodied CO_2, or 2.45×10^8 kg, was avoidable if these buildings had been repaired.

Although this provides a rough indicator as to the environmental impacts of demolishing relatively lightly damaged buildings following earthquakes, the significant savings in embodied CO_2 provide a preliminary justification for incorporating environmental impacts into demolish/repair decisions following earthquakes. Furthermore, the above assessment underestimates the environmental impacts of building demolitions as it has only considered the embodied CO_2 and energy in building materials. Further work by the authors will seek to include the impacts from the Construction and Waste Management Stages (Modules 2 and 3 of Fig. 3.7) in estimating the total environmental impacts of building demolitions after earthquakes.

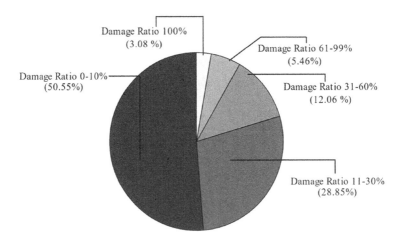

Fig. 3.11 Percentage of total embodied carbon and energy based on damage ratio

3.4 Summary and Conclusions

Past studies which evaluated the complex factors that contributed to the decisions to demolish or repair buildings in Christchurch following the Canterbury Earthquakes were summarised. These studies identified a number of quantitative and qualitative factors which led to the high demolition rate including building parameters and structural damage, as well as high insurance rates, and changes and confusion in local building legislation. The review of these studies revealed that environmental impacts were neglected in the repair/demolish decision framework following the Canterbury Earthquakes. To begin to provide a justification that environmental impacts should be included in the demolish/repair framework following earthquakes, a rough quantitative assessment of the environmental impacts of building demolitions in the Christchurch CBD was conducted in terms of the embodied CO_2 and energy in the building materials assuming a like-for-like replacement of materials that were demolished. Despite ignoring the wider impacts of all complex factors that influence the environmental impacts, this initial assessment revealed high environmental impacts associated with building demolitions following earthquakes. A more in-depth analysis of the data revealed that a large amount of embodied CO_2 in the Christchurch demolitions (~80% in the building set used here) was from buildings which were relatively lightly damaged following the event. This highlights the importance of incorporating environmental impacts into the decision to demolish or repair a building following an event, particularly if the building is relatively undamaged. The work presented here was an initial study which will be extended to include wider environmental impacts including construction methods, repair (as an alternative to demolition), waste management (in the case of demolition) including the economic viability of encouraging the diversion of deconstructed materials away from landfills.

References

Bevere L, Grollimund B (2012) Lessons from recent major earthquakes | Swiss Re

BRANZ (2019) Whole-building whole-of-life framework [WWW Document]. BRANZ CO2NSTRUCT. https://www.branz.co.nz/environment-zero-carbon-research/framework/branz-co2nstruct/

Carbon Leadership Forum (2019) Estimates of embodied carbon for mechanical, electrical, plumbing and tenant improvements

Chang SE, Taylor JE, Elwood KJ, Seville E, Brunsdon D, Gartner M (2014) Urban disaster recovery in Christchurch: the central business district cordon and other critical decisions. Earthq Spectra 30:513–532. https://doi.org/10.1193/022413EQS050M

Department of Building and Housing (2011) Compliance document for New Zealand building code. Wellington, New Zealand

Dowdell D, Berg, B (2016) New Zealand wholebuilding whole-of-life framework: An overview (No. SR349). BRANZ, Wellington, New Zealand

FEMA (2012) FEMA P-58: Seismic performance assessment of buildings. Federal Emergency Management Agency, Washington D.C

Isaacs N, Camilleri M, Burrough L, Pollard A, Saville-Smith K, Fraser R, Rossouw P, Jowett J (2010) Energy use in New Zealand households (No. SR 221). BRANZ, Wellington, New Zealand

Kam WY, Pampanin S (2011) The seismic performance of RC buildings in the 22 February 2011 Christchurch Earthquake. Struct Concr 12:223–233. https://doi.org/10.1002/suco.201100044

Kim JH (2015) Quantitative analysis of factors influencing post-earthquake decisions on concrete buildings in Christchurch. University of British Columbia, Vancouver, British Columbia, New Zealand

Kim JJ, Elwood KJ, Marquis F, Chang SE (2017) Factors influencing post-earthquake decisions on buildings in Christchurch. New Zealand. Earthq. Spectra 33:623–640. https://doi.org/10.1193/072516EQS120M

Marder K, Elwood KJ, Motter CJ, Clifton GC (2020) Post-earthquake assessment of moderately damaged reinforced concrete plastic hinges. Earthq Spectra 36:299–321. https://doi.org/10.1177/8755293019878192

Marquis F (2015) A Framework for understanding post-earthquake decisions on multi-storey concrete buildings in Christchurch. University of British Columbia, Vancouver, British Columbia, New Zealand

Marquis F, Kim JJ, Elwood KJ, Chang SE (2017) Understanding post-earthquake decisions on multi-storey concrete buildings in Christchurch, New Zealand. Bull Earthq Eng 15:731–758. https://doi.org/10.1007/s10518-015-9772-8

MBIE (2020) Building for Climate Change: Transforming the building and construction sector to reduce emissions and improve climate resilience. Ministry of Business, Innovation and Employment, Wellington, New Zealand

MBIE (2018) The Seismic assessment of existing buildings: technical guidlines for engineering assessments. Wellington, New Zealand

Ministry for the Environment (2019) Measuring emissions: a guide for organisations. 2019 Summary of emission factors. Ministry for the Environment, Wellington, New Zealand

New Zealand Supreme Court (2014) New Zealand Supreme Court (NZSC) (2014) University of Canterbury v Insurance Council of New Zealand

Parker M, Steenkamp D (2012) The economic impact of the Canterbury earthquakes. Reserve Band N Z Bull 75

Potter SH, Becker, JS, Johnston DM, Rossiter KP (2015) An overview of the impacts of the 2010–2011 Canterbury earthquakes. Int J Disaster Risk Reduct, The 2010–2011 Canterbury Earthquake Sequence: Personal, Social, Governance and Environmental Consequences 14, 6–14. https://doi.org/https://doi.org/10.1016/j.ijdrr.2015.01.014

Chapter 4
Damage Assessment in Italy, and Experiences After Recent Earthquakes on Reparability and Repair Costs

M. Di Ludovico, G. De Martino, A. Prota, G. Manfredi, and M. Dolce

Abstract Recent devastating earthquakes outlined the importance of quantifying losses and the amount of resources needed for the reconstruction process. The restoration of public or residential buildings in the aftermath of the seismic event may significantly affect national economy. This remarks the primary role and crucial need of having accurate predictions of direct and indirect costs for reconstruction in order to plan effective risk mitigation strategies and perform reliable loss scenarios. The recent Italian seismic events have been a unique occasion to collect observational data on existing buildings. The present work, based on the Italian experience of recent earthquakes, aims at discussing the main aspects related to the damage assessment of residential buildings and reconstruction models together with the huge amount of data collected in the reconstruction processes. In particular, an in-depth analysis of the data provided by the reconstruction process of 2009 L'Aquila earthquake is reported focussing on repair and strengthening intervention costs as a function of the empirical damage, repairability issues, and assistance to population costs. The data are discussed separately for reinforced concrete and masonry residential buildings and refers about 10,100 buildings located Outside Historical Centres (OHC) and Inside Historical Centres (IHC). Finally, the criteria adopted for the definition of the building seismic risk classes at the base of the Italian guidelines for seismic risk classification of constructions are presented together with recent policies adopted in Italy in terms of fiscal deduction for strengthening interventions on private residential buildings.

M. Di Ludovico (✉) · G. De Martino · A. Prota · G. Manfredi · M. Dolce
Italian Civil Protection Department, Presidency of the Council of Ministers, Via Ulpiano 14, Rome 80125, Italy
e-mail: marco.diludovico@unina.it

© The Author(s) 2021
S. Akkar et al. (eds.), *Advances in Assessment and Modeling of Earthquake Loss*,
Springer Tracts in Civil Engineering,
https://doi.org/10.1007/978-3-030-68813-4_4

4.1   Introduction

Existing structures often exhibit poor seismic performance as demonstrated by the diffuse damage and numerous collapse, either partial or total, surveyed in the aftermaths of moderate-to-high magnitude strong motions worldwide; damage provided by earthquakes is a concern for a society as a whole in terms of loss of life and direct and indirect costs.

Italy has experienced more than 60 destructive earthquakes over the past two centuries and starting from the devastating earthquake of Belice in 1968, the death toll has been about 5,000, corresponding to approximately 100 deaths/year. In addition, direct costs and indirect costs have dramatically affected the country's economy. The direct costs only related to the emergency management and reconstruction process in Italy between 1968 and 1998 were estimated to exceed €100 billion (by the 2005 euro equivalent), mainly related to the earthquakes in Belice (1968), Friuli (1976), Irpinia (1980) and Umbria-Marche (1997), (Severino and Di Pasquale 2002). These costs are considerably increased if due allowances are made for the seismic events of the last 15 years, including events in Molise (2002), L'Aquila (2009), Emilia Romagna (2012) and Central Italy (2016–2017). Indeed, the L'Aquila earthquake left nearly 70,000 homeless, the Emilia earthquake strongly impacted on productivity of primary importance for the local and national economy, and the central Italy earthquake highlighted the cumulative effects of a seismic sequence on the damage to buildings and relevant losses.

A proper quantification of lives and monetary losses as well as of time to recover the buildings' functionality is of paramount importance to give indications to decision makers for establishing seismic risk mitigation policies, and to insurance companies to value sound insurance premium for existing building in the seismic prone areas.

To this aim, it is fundamental to collect post-earthquake data regarding the usability of buildings, the type and extent of damage on structural and non-structural members, the ordinances issued to regulate the reconstruction stages and the relevant costs and time to be completed.

The data on post-earthquake surveys carried out after last 50 years devastating earthquakes in Italy have been recently collected in a wide database reported in a web-based platform named Da.D.O. (Database of Observed Damage), (Dolce et al. 2019); it reports data on about 320,000 buildings inspected after earthquakes that stroke several Italian regions from 1976 to 2012. Since the Umbria-Marche 1997 seismic event, the damage and usability assessment of buildings has been made by the first level AeDES survey form, (Baggio et al. 2007). The form represents a rapid tool to assess the damage and usability based on the visual in situ inspection of the building. The form refers to the minimum structural unit with a significant impact on the people safety and reports data on damage level and extent on structural and non-structural members evaluated by teams of experts in seismic engineering. Similarly, to other forms used all around the world (e.g. Japan (Goretti and Inukai 2002), U.S. (ATC 2005), New Zealand (NZSEE 2009)), the main goal is to assess usability categories. For example, according to ATC (2005), a building is tagged "Green" for unrestricted access, "Yellow", for restricted access, and "Red" for no access, while the AeDES form leads to six usability categories: A. Usable buildings; B. Building usable only after short term countermeasures; C. Partially usable building; D. Building to be re-inspected; E. Unusable building; F. Unusable building for external risk.

Once usability of buildings has been evaluated, the reconstruction process can be managed. The models of post-earthquake emergency management and reconstruction used in Italy since the 1968 Belice earthquake have all been based on ensuring fair public coverage of the costs required to repair the earthquake damage while different economic thresholds have been defined for local or global strengthening interventions. The technical and administrative policies for the implementation of the reconstruction have been refined over time to allow for the experience of previous earthquakes and for the improvement of technical and scientific knowledge.

The present work, based on the experience of recent 2009 L'Aquila earthquake for which it was possible to collect a huge amount of data, aims at discussing the main aspects related to the damage assessment of buildings as well as the aspects related to direct and indirect costs for reconstruction (i.e. repair and strengthening intervention costs, repairability, and assistance to population costs). The data discussed herein have been used in Italy to define a document specifically developed for the seismic risk classification of existing buildings approved in February 2017 by the Consiglio Supe-riore dei Lavori Pubblici, (Ministry Decree no. 2017) defining the technical principles for exploiting tax deductions with respect to seismic strengthening interventions (the so-called "*Sismabonus*").

4.2   The 2009 L'Aquila Earthquake Experience

The 2009 L'Aquila (Abruzzi Region) earthquake affected 57 municipalities with MCS intensity greater than or equal to VI (Dolce 2010) and caused extensive damage to public and private structures, to artistic and cultural heritage of L'Aquila and rele-vant provinces with a huge number of homeless people. The maximum number of people assisted, in the days immediately following the main event on 6 April 2009, was 67,459 people, allocated in 171 tent camps and in hotels or other accommodation facilities located mostly on the Adriatic coast. In the immediacy of the event, tempo-rary accommodation was realized to host population and essential public functions, such as schools (Decree of the head of the USRC February 06 2014; Decreto Legge 19 maggio 2020).

Once the state of emergency was declared, the damage and usability assessment of the private and public buildings, under a central coordination of the Civil Protec-tion Department, was activated in order to determine whether they could be safely used. Since the State Government intended to provide a considerable public financial support to the reconstruction process, specific policies were adopted in the municipal-ities that experienced a macro-seismic intensity greater than or equal to VI, according to the MCS Scale, Mercalli-Cancani-Sieberg (Grünthal 1998): the so called "Crater" included L'Aquila and other 56 municipalities. The reconstruction process involved two different models: the "analytical model" issued in the first stage of reconstruction from 2009 to 2013 for private buildings outside the historical centres (OHC) of Crater, and the "parametric model" adopted in a second stage for private buildings inside the historical centres (IHC) of L'Aquila and other Crater municipalities. The analytical model involved two different reconstruction steps, "light damage" reconstruction related to B or C rating residential buildings and "heavy damage" reconstruction related to E rating residential buildings.

The reconstruction process after the L'Aquila earthquake (2009) have offered a unique opportunity to collect and monitor data on a large scale.

Efforts to analyze these data have resulted in a unique database of 5,775 records related to residential buildings OHC (Annex to OPCM no. 3779 2009; Annex to OPCM no. 3790 2009; Di Ludovico et al. 2017a, b; OPCM no. 3779 2009; OPCM no. 3790 2009; OPCM no. 3881 2010), 1,170 records related to residential buildings IHC (to be increased in the future because the reconstruction process is still ongoing) and to 53,968 displaced people assisted in the emergency and reconstruction stages, (Mannella et al. 2017).

The analysis of the data collected with reference to the analytical and parametric models are presented in the next sections. In particular, the data collected on buildings located OHC and IHC are presented and discussed focusing on the main statistics related buildings' type, damage, cost data for repair and strengthening interventions as well as reparability issues.

4.3   The Reconstruction of Residential Building Outside Historical Centers (OHC)

The reconstruction process of residential buildings outside the historical centres (OHC) damaged by the L'Aquila earthquake was calibrated on the basis of damage and usability assessment of each private building. The first stage of the reconstruction process involved B or C rating buildings, the so-called "light damage" reconstruction, while in a second stage the recovery involved E rating buildings, the so-called "heavy damage" reconstruction. This to differentiate the stages of the reconstruction process as a function of the observed damages on the vertical structures as issued in the specific post-earthquake ordinances, (Annex to OPCM no. 3779 2009; Annex to OPCM no. 3790 2009; OPCM no. 3779 2009; OPCM no. 3790 2009; OPCM no. 3881 2010). The distinction in two stages enabled rapid re-occupancy of slightly damaged buildings, thus significantly reducing public costs incurred in housing the homeless. Details about the data related to both "light damage" and "heavy damage" reconstruction are reported in (Di Ludovico et al. 2017a, b).

According to the ordinances specifically issued for the reconstruction of damaged buildings, the repair costs to restore original condition of damaged structural or non-structural members were fully covered by the public grant. In addition, according to the *building back better*" principle, strengthening intervention costs were also covered by the Government in order to reduce the vulnerability of repaired buildings, together with structural and geotechnical tests and energy efficiency upgrade. A suitable technical documentation, carried out by practitioners engaged by owners, was required to illustrate the damage, the design of repair and strengthening interventions, and to quantify the government financial support required (i.e. application for funding). The public grant was released once an administrative, technical and economical check was made by a proper commission, called "*Filiera*" (i.e. an Italian word to indicate a supply chain mechanism). The Filiera activity allowed to collect a database containing technical and economic information on 5,775 residential buildings OHC, of which 4,855 buildings (3,546 (i.e. 62%) B or C rating buildings and 2,211 (i.e. 38%) E rating buildings) of L'Aquila municipalities and 920 buildings of other municipalities of the Abruzzi Region (660 (i.e. 72%) B or C rating buildings and 260 (i.e. 28%) E rating buildings.

The total amount of public grant allocated for 4,855 buildings (2,904 B or C and 1,951 E rating) residential buildings OHC of L'Aquila can be estimated of the order of 2.6 billion euros: 0.5 billion euros for B or C rating residential buildings, and 2.1 billion euros for E rating residential buildings. Out of 2.6 billion euros, 1.3 billion euros, involved repair interventions while 0,7 and 0,6 billion euros involved seismic strengthening and demolition/reconstruction interventions, respectively.

In the application for funding, E rating buildings were further classified in three funding-classes: class E–B, including buildings with a high non-structural risk that sustained medium structural damage (where a local strengthening strategy may solve most of the structural weakness); E including buildings with severe structural damage; and class E_{dem}, including buildings that needed to be demolished because of dangerous structural weaknesses, a high residual drift, local or global collapse, or a lack of economic value of required repair or strengthening interventions compared to the costs of demolition and reconstruction (Di Ludovico et al. 2017a).

The maximum grant for strengthening interventions was established as a function of the usability rating of buildings and relevant funding classes, as shown in Fig. 4.1.

The grants allocated for demolition and reconstruction involved 541 buildings out of the 2,211 E rating buildings (i.e. about 24% of the dataset): 539 in L'Aquila municipality and 2 in other municipalities. For buildings in L'Aquila municipality, a public grant computed on the basis of *forfait* unit costs was requested for 39 buildings: €500/m^2 for 17 buildings, and €750/m^2 for 22 buildings. As per the remaining 500 buildings, the reparability (and seismic strengthening) resulted to be an option not viable for several reasons. The grant for demolition and reconstruction was computed: on the basis of economic convenience for 421 buildings; without economic assessment for 44 masonry buildings partially collapsed (more than 25% in volume); for 34 R.C. buildings with average compressive cylindrical strength fcm <8 MPa; and for 1 R.C. building with more than 50% of story's columns with a drift greater than 1.5%. The mean public grant resulted: €1,192/m^2.

A detailed description of the reconstruction policy, the regulation and an overview of the database of 5,775 residential buildings damaged by the L'Aquila earthquake is reported in (Di Ludovico et al. 2017a, b).

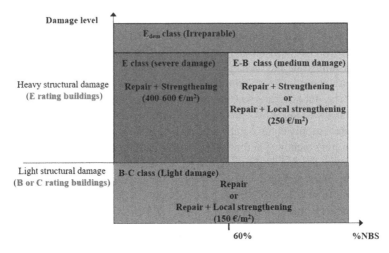

Fig. 4.1 Repair/Strengthening criteria: policies after 2009 L'Aquila earthquake (NBS = New Building Standard)

In the next sections, the analysis focuses on the data related to empirical damage (derived from the AeDES forms), intervention costs (derived from the applications for funding), and time and costs for population assistance (derived from municipalities offices).

4.3.1 Damage and Repair Costs

The data of damaged buildings OHC involves 5,775 buildings; out of those buildings, 95% of the database concerns RC or masonry buildings (49% RC and 46% masonry, respectively), while the remaining 5% involves buildings with a mixed structural type (i.e. comprising RC and masonry structural members), steel structure or other types. The number of buildings for each structural type and their usability rating is summarized in Table 4.1. It shows that RC is the most common structural type in B

Table 4.1 No. of buildings in each structural types and usability rating class

Building stock	Structural type	No. of buildings	Damage	Usability rating	No. of buildings
5,775	Masonry	2,673	Light	B or C	1,580
			Heavy	E	1,093
	RC	2,797	Light	B or C	1,738
			Heavy	E	1,059
	other types	305	Light	B or C	246
			Heavy	E	59

or C rating buildings; by contrast, masonry is the most common structural type in the case of E rating buildings.

The repair costs monitored during the reconstruction process of private residential buildings OHC damaged by L'Aquila earthquake (2009) were collected for a subset of 3,992 buildings (i.e. 2,512 RC buildings and 1,480 masonry buildings). The repair costs included in the L'Aquila reconstruction database are inclusive of: building safety measures; demolition and removal, including transportation costs and landfill disposal; repair interventions; repair and finishing works relevant to strengthening interventions; the testing of facilities; technical works for health and hygiene improvement; technical works to improve facilities; construction and safety costs; fees for the design and technical assistance of practitioners; and furniture moving. They do not include value added tax (VAT). Table 4.2. summarizes the mean repair costs normalized by the overall building gross surface area (i.e. unit costs expressed in €/m²) as a function of funding classes and structural types. For the sake of simplicity, the four-funding class have been also identified in four damage classes: light, medium, severe, irreparable.

By associating these costs data with information related to the empirical damage experienced on structural and non structural members of such buildings, it was possible to define values of $\%c_j$, with j from 1 to 5 (defined as a percentage of the reconstruction cost of new building established equal to €1,350/m² at national level). The $\%c_j$ are associated to each global Damage State, defined in compliance with the metric introduced in EMS98 (Grünthal 1998), (i.e. five global Damage States from DS1 to DS5), see Table 4.3.

In particular, DS_j Global damage grades have been obtained by means of suitable conversion matrices of empirical damage data collected in the AeDES forms defined according to (Del Gaudio et al. 2017) for RC buildings and (Dolce et al. 2019)

Table 4.2 Mean unit costs related to RC and masonry buildings in L'Aquila

Damage	Funding class	Type of Structure	No. of buildings	Repair costs
	(–)	(–)	(–)	(€/m²)
Light	B or C	RC	1,598	183.76
		Masonry	899	216.81
		All	2,497	195.66
Medium	E–B	RC	200	342.35
		Masonry	44	268.29
		All	244	328.99
Severe	E	RC	447	532.90
		Masonry	313	447.85
		All	760	497.87
Irreparabile	E$_{dem}$	RC	267	1,213.40
		Masonry	224	1,169.85
		All	491	1,192.00

Table 4.3 Percentage cost of repair or replacement as a function of global Damage State

	Global Damage State, DS_j				
	DS1	DS2	DS3	DS4	DS5
Percentage of reconstruction cost of new building, $\%c_j$	2%	10%	30%	60%	100%

for masonry buildings (see Fig. 4.2). The data of Table 4.3. are reported in the document assessing the Italian national seismic risk recently edited by the National Civil Protection Department (2018).

Furthermore, the data related to repair costs have been analyzed in detail in (Del Vecchio et al. 2020), in order to evaluate the influence of structural and non-structural members and drift/acceleration sensitive members on building repair costs, *BRC*, of reinforced concrete buildings. The *BRC* is obtained excluding from the repair costs computed by practitioners, general costs for construction field installation, safety measures, professional fees, external works and repair costs related to structural strengthening intervention.

The study points out that, for the selected subset of 120 buildings, the repair costs related to partitions and infills ranges from 43 to 58% of *BRC*. Because in the Mediterranean construction system, plumbing and electrical systems are commonly incorporated in hollow clay brick partitions and infills, by adding the repair costs of these components as well as of windows and doors and enclosure systems, the repair cost ratio rise to 81–89% of the *BRC*. Furthermore, the analyses outline that 63–70% of the *BRC* concerns the repair of drift-sensitive components, while 15–21% relates

Empirical damage (Aedes form)		Global Damage State, DSj			
		Dolce et al, 2019		Del Gaudio et al, 2017	
Severity	**Extension**	**Vertical Structure**		**Vertical Structure**	**Infill partitions**
Null	-	DS0		DS0	DS0
D1 Light	<1/3	DS1		DS1	DS1
	1/3-2/3	DS1		DS1	DS1
	>2/3	DS1		DS1	DS1
D2-D3 Medium-Heavy	<1/3	DS2		DS2	DS2
	1/3-2/3	DS3		DS3	DS2
	>2/3	DS3		DS3	DS2
D4-D5 Very Heavy - Collapse	<1/3	DS3 ($+k_{D2-D3}<1/3$)	DS3 ($+k_{D2-D3}>1/3$)	DS4	DS3
	1/3-2/3	DS4 ($+k_{D2-D3}<1/3$)	DS5 ($+k_{D2-D3}>1/3$)	DS4	DS3
	>2/3	DS5		DS5	DS3

Fig. 4.2 Conversion matrices of empirical damage data collected in the AeDES forms

to the repair of those that are acceleration-sensitive (i.e. roofs and chimneys, sanitary and other equipment, floor finishes).

4.3.2   Strengthening Intervention, Structural/Geotechnical Tests and Energy Efficiency Costs

The technical documentation provided by practitioners to the Filiera allowed to collect the design drawings of the repair and, in several cases, strengthening interventions. To support the engineers involved in the L'Aquila reconstruction process, in August 2009, DPC and ReLUIS published a proper guideline "*Guidelines for Repair and Local Strengthening of Structural and Non-Structural Members*" (Civil Protection Department (DPC) and Laboratories University Network of Seismic Engineering (ReLUIS) 2011), to drive practitioners involved in the reconstruction process.

In case of B or C rating buildings according to the AeDES classification, local strengthening solutions were adopted and mainly involved the use of composite materials (i.e. FRP), because they appeared very effective to increase the capacity of vulnerable elements (beam column joints and short columns in RC buildings, and wall connections in masonry ones) without significantly affecting the building global mass and stiffness. Furthermore, very common was the use of steel plates or ties for RC and masonry buildings, respectively.

In case of E rating buildings, a global strengthening strategy was adopted to significantly increase the buildings structural capacity; a safety threshold at ultimate limit state equal to 60% of New Building Standard (%NBS = 60%) was mandatory to have access to the public grants for the reconstruction of severely damaged buildings (see Di Ludovico et al. 2017a; OPCM no. 3779 2009) for more details). The strategy to improve the seismic capacity of existing buildings commonly involved the use of several techniques, and, in many cases, the combination of traditional strengthening systems with innovative ones. Note that 59 buildings with severe damage were retrofitted by using base isolation and 13 by using energy dissipation bracing systems. In case of buildings with severe damage (E rating buildings) energy efficiency interventions were also covered by public grant.

The mean costs of local or global strengthening, structural and geotechnical tests and energy efficiency upgrade on RC or masonry structures are summarized in Table 4.4.

The technical documentation provided by practitioners included the seismic capacity assessment of the building in the ante and post-operam configuration to check the initial structural capacity and the attainment of at least %NBS = 60%. The seismic structural safety has been assessed as the ratio between demand and capacity peak ground acceleration. In order to evaluate the effectiveness of the seismic strengthening intervention, similarly to what has been done for strengthening of existing undamaged buildings all over Italy (Dolce et al. 2019), Fig. 4.3.

Table 4.4 Mean unit strengthening intervention, structural/geotechnical tests and energy efficiency costs related to RC and masonry buildings in L'Aquila (Di Ludovico et al. 2017a, b)

Damage	Usability rating	Type of Structure	No. of buildings	Strength. costs	Structural and geotech. tests	Energy effic. upgrade
(–)	(–)	(–)	(–)	($€/m^2$)	($€/m^2$)	($€/m^2$)
Light	B or C	RC	1598	33.9	–	–
		Masonry	899	68.32	–	–
		All	2497	46.29		
Medium	E–B	RC	200	139.01	3.99	39.9
		Masonry	44	143.7	4.27	34.3
		All	244	139.86	4.04	38.89
Severe	E	RC	447	309.24	7.84	75.82
		Masonry	313	320.13	10.23	59.08
		All	760	313.72	8.82	68.93

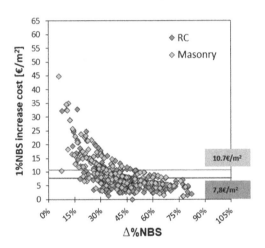

Fig. 4.3 The unit strengthening costs per point of seismic strengthening as a function of Δ%NBS

summarizes the unit strengthening costs per point of safety gain due to the seismic upgrading (Δ%NBS) in function of the increase of %NBS, Δ%NBS.

This data shows that the cost for per square meter per %point increase of Δ%NBS ranged between 1.2 and 44.7 $€/m^2$ and on average resulted equal to €7.8 m^2 and €10.7/m^2 for RC and masonry building in E funding class. In details, the higher the Δ%NBS is, the lower is the mean unit cost to increase the %NBS by one percentage unit, as also shown in (Dolce et al. 2019). The data of Fig. 4.3. may allow to easily predict the cost of the strengthening intervention to reduce the vulnerability of existing buildings. However, some caution should be taken because they refer to data derived from L'Aquila 2009 post-earthquake reconstruction process. Furthermore, it was mandatory to achieve by strengthening interventions %NBS values between 60 and 80% to have access to the public grant in the L'Aquila earthquake reconstruction process.

4.3.3 Population Assistance: Accommodation Costs

The 2009 L'Aquila earthquake left 67,459 homeless people, an impressive number if compared with the resident population in the Crater area at the time of the earthquake: 68,503 in L'Aquila municipality and 71,081 in other municipalities for a total of 139,584 residents. In the immediate post-event 35,690 people were accommodated in 171 camps with 5957 tents and 31,769 in hotels and private homes, (Presidency of Council of Ministers, Civil Protection Department 2010).

Then, to move people from tents and to provide short-term and long-term accommodations to homeless people, a financial assistance was given by the Italian government for several solutions: (i) accommodation in hotels or public structures, (ii) self-accommodations grant, s.a.g., in the following; and (iii) permanent structures as the common wooden-house units (i.e. M.A.P.—Temporary Inhabitable Modules) or the new solution, so-called "C.A.S.E. project", Anti-seismic, Sustainable and Ecologically Compatible Housing Complexes.

People in slightly damaged buildings (B or C usability rating according to AeDES classification) were hosted in (i) and (ii) solutions while the latter solutions (M.A.P. modules and buildings of the C.A.S.E project) were mainly addressed to manage the long-term recovery. They hosted people who lived before the earthquake in buildings severely damaged by the quake (E usability rating according to AeDES classification) or located within the perimeters of the so-called "Red Zone", a restricted area of the town with buildings prone to collapse.

M.A.P. modules and buildings of the C.A.S.E project were realized to host a maximum number of 18,000 people in the L'Aquila municipality, (Dolce and Di Bucci 2017).

The number of people assisted in each accommodation solution in each semester from December 2009 to December 2016 is depicted in Fig. 4.4a. The reconstruction policy based on promoting "light damage" reconstruction prior to "heavy damage" reconstruction allowed 21,960 people returning home after one year and eight months from the earthquake and 43,134 (i.e. about 80% of people needing assistance in December 2009) after seven years and eight months from the earthquake (42,408 after six years and eight months from the earthquake). The Filiera activity ended in 2013, but 2015 may be considered the end of the returning home trend of such stage, see Fig. 4.4b.

The costs for people assistance resulted equal to about 0.32 and 0.24 billion euros for accommodation in hotels or public structures and s.a.g., respectively. Furthermore, the costs for construction of M.A.P. modules and buildings of the C.A.S.E project were 0.12 and 0.85 billion euros, respectively. Thus, by summing such costs, a total amount of about 1.5 billion euros can be estimated as accommodation costs;

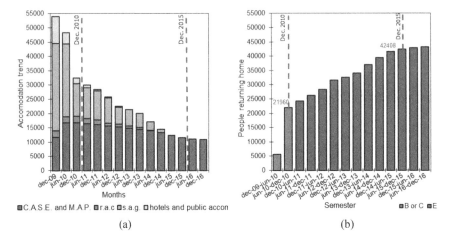

Fig. 4.4 Accommodation trend according to assistance type (**a**) and people returning at home trend (**b**)

however, note that 0.97 billion euros is the full cost of M.A.P. and C.A.S.E. accommodations which are still hosting homeless and will remain usable in the future for alternative purposes and occupation. The data related to population assistance timing and costs strongly highlights the impact of indirect costs in a proper evaluation of post-earthquake losses.

The reconstruction of buildings inside the historical centres was obviously affected by the difficulties related to the design interventions on old masonry building aggregates with a cultural and architectural heritage value and relevant need to preserve their artistic and architectonic assets. The reconstruction of buildings inside historical centers of L'Aquila town and surrounding villages is currently ongoing and described in the following.

4.4 Reconstruction of Residential Buildings Inside Historical Centers (IHC)

In the reconstruction process of the historical centres damaged by the 2009 L'Aquila earthquake, it was introduced a simplified procedure to easily take into account both the structural complexity of the historical centres and the territorial peculiarities (i.e. materials and construction techniques and valuable architectural components). Indeed, the historical centres are mostly made of Building Aggregates (BA). They consist of portions with homogeneous characteristics and with low or without mutual seismic dynamic interactions: the so-called Aggregate Minimum Unit—AMU. Each AMU can be made of one or more buildings (B) with the same or with different usability ratings.

Buildings Inside Historical Centers (IHC) presents structural and non-structural elements typical of historical architecture, such as contrast arcs, vaults and loggias, whose behaviour and mutual interaction make their seismic response difficult to simulate and predict in numerical analyses. Structural and architectural interventions carried out on the original structures over time, makes even more difficult the design of repair and strengthening interventions. Therefore, the implementation of a parametric model was essential to determine the maximum allowable public grant to restore the usability of damaged buildings and to increase their seismic safety.

According to D.P.C.M 54/2013 (February 4 2013), the analytical model was replaced by the parametric one; it introduced two different special reconstruction Offices to manage such reconstruction stage: (i) the Special Reconstruction Office of L'Aquila, (USRA), for the reconstruction process IHC of L'Aquila; (ii) the Special Reconstruction Office of the Crater Municipalities (USRC) for the reconstruction process of IHC of other municipalities.

Each office developed a parametric model to manage the reconstruction process and to define the maximum public grant to repair and strengthen the damaged buildings (Decree of the head of the USRA January 21 2013; Decree of the head of the USRC February 06 2014; Fico et al. 2017). According to the parametric models, the grant was established by means of two main steps: (i) definition of funding amount threshold (namely Allowable Grant, AG) by means of parametric costs established as a function of building vulnerability class and damage assessed through AeDES form; (ii) definition of repair and strengthening intervention and relevant costs carried out by practitioners engaged by owners. Both models fully cover not only the repair and strengthening costs to restore the usability buildings but also interventions to preserve the cultural and architectural heritage value of IHC buildings.

The technical documentation submitted by practitioners to Special Offices refers to single AMUs, which may coincide with a single building (B) or with the entire aggregate (BA) or with the portion of the aggregate composed of one or more buildings.

To date the reconstruction process of IHC residential buildings is still ongoing. At the end of 2019, 3,938 applications for funding for repair and strengthening interventions on AMUs were submitted for the reconstruction process IHC: 1,581 to USRA, and 2,357 to USRC. The Special Offices approved 1,170 applications for funding (526 by USRA and 644 by USRC). The total amount of public grant allocated for IHC residential buildings until December 2019 was about €1,9 billion (€1,15 billion by USRA, L'Aquila municipality, and 0,75 billion by USRC, Crater municipalities).

The historical centres are mainly characterised by masonry buildings. Thus, the data on masonry buildings of OHC (analytical model) are only used herein for comparison with those provided by the parametric model. However, note that masonry buildings IHC are mainly characterised by rubble masonry, often made of materials with low mechanical properties and lacking efficient earthquake-resistant structural details.

An analysis containing data related to 526 and 644 AMU corresponding to 1,472 and 2,855 buildings located in L'Aquila municipality and in other Crater municipalities, respectively, is herein discussed. Thus, a dataset of 5,763 buildings is herein analysed.

The AMUs are generally made by one or more buildings and mostly consists of two, three or four buildings. On average, the number of buildings per AMU resulted to be 2.8 and 4.4 for L'Aquila and Crater municipalities, respectively.

The reconstruction policy adopted in the IHC established strategies of urban reconstruction compatible with the different historical, cultural, natural, morphological and aesthetic value levels of historical centres with the aim of recovering the pre-existing cultural heritage values. To this aim, different categories have been defined for buildings with: (i) historic-architectural valuable elements; (ii) landscape interest; (iii) specific heritage protection provisions (the so-called building of cultural interest). Buildings not included in previous categories are defined "ordinary". Note that, according to such assumption, the buildings OHC were all identified as ordinary.

An inventory of valuable elements and percentage of grant increase was also defined in order to support the preservation of buildings IHC (Fico et al. 2017). Grant increase was introduced within the following maximum limits: 60% of the allowable grant AG for building with historic-architectural valuable elements; 100% of the AG for building with landscape interest and of cultural interest. Figure 4.5. reports the percentage distribution of building categories OHC (Filiera) and IHC (for L'Aquila municipality, managed by USRA, and Crater municipalities, managed by USRC).

In order to better understand the influence of valuable elements of IHC buildings with respect to OHC buildings, Table 4.5. reports, for a subset of buildings, the unit costs of repair and strengthening interventions as a function of the building category (i.e. ordinary buildings, buildings with historic-architectural valuable elements, building with landscape interest building of cultural interest) and usability ratings. Number of buildings are reported in square brackets.

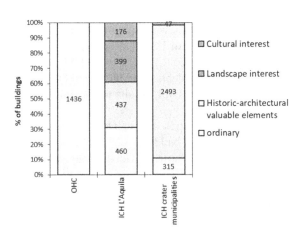

Fig. 4.5 Distribution of buildings categories OHC and IHC

Table 4.5 Unit repair and strengthening costs as a function of building categories (number of buildings in square brackets)

Building category	B or C rating			E rating		
	OHC	IHC Crater municipalities	IHC L'Aquila	OHC	IHC Crater municipalities	IHC L'Aquila
Ordinary	285.13 [899]	323.75 [6]	–	767.98 [313]	944.0 [8]	866.26 [95]
Valuable elements	–	375.47 [46]	–	–	1,032.61 [409]	1,185.36 [253]
Landscape interest	–	–	–	–	–	1,643.70 [290]
Cultural interest	–	–	–	–	1,753.33 [18]	2,229.71 [155]
All	285.13 [899]	369.01 [52]	–	767.98 [313]	1053.71 [435]	1,518.87 [793]

The analysis shows that the repair and strengthening costs related to grant for OHC buildings resulted on average lower than those for IHC buildings, independently from usability rating. The reason of such cost increase is clearly related to the extra costs needed to preserve, restore or repair valuable elements on buildings IHC.

4.5   Seismic Risk Classification of Constructions in Italy

The earthquakes experienced in Italy in recent years clearly showed that the loss of lives, monetary-losses and the resources/time for recovery of building functionality are no more acceptable in the future. Thus, Italy has been the first country in Europe adopting a methodology explicitly correlating the seismic risk, the performance of structural and non-structural members and the expected losses. On February 2017, the "Guidelines for the seismic risk classification of the constructions" have been approved by the Consiglio Superiore dei Lavori Pubblici (Ministry Decree no. 2017). They define the technical principles for exploiting tax deductions with respect to seismic strengthening interventions on existing private buildings (the so-called "Sismabonus"). The guidelines define eight-risk classes (A+, A, B, C, D, E, F, G). The seismic risk classes of buildings and the class changes due to the strengthening interventions can be assessed using the principles included in the guidelines. The risk class can be determined based on a very simple procedure allowing practitioners to deal with concepts such as the expected annual losses (EAL) and the repair costs expressed as a percentage of the reconstruction cost of new building ($\%c_j$), (Cosenza et al. 2018). In the latter case the guidelines define seven seismic risk classes in function of the building safety index at the ultimate limit state (%NBS). In particular, the seismic risk class is defined as the minimum one of the class related to the building safety index at the ultimate limit state (%NBS) and the class related to EAL. The

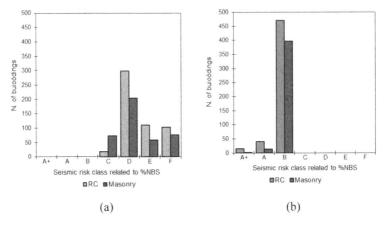

(a)   (b)

Fig. 4.6 Seismic risk class related to %NBS for RC and masonry buildings severely damaged by 2009 L'Aquila earthquake in the original *ante-operam* configuration (**a**) and in the *post-operam* configuration (**b**)

latter class depends on the area under the curve of the expected losses, which has been also calibrated by using data costs monitored in the L'Aquila reconstruction process.

The safety index class expressed by computing the %NBS index for 527 and 410 respectively reinforced concrete and masonry buildings severely damaged by the L'Aquila earthquake is reported in Fig. 4.6. The %NBS index has been computed by practitioners and it refers to undamaged structure (i.e. as before the earthquake) and to the strengthened structure according to designs (Di Ludovico et al. 2017a; OPCM no. 3779 2009) and then converted in one of the seven seismic risk classes defined by the Guidelines.

Figure 4.6a shows that no buildings, in the considered sample of damaged ones, belong to classes safer than C seismic risk class in the original *ante-operam* configuration, while the strengthening interventions allowed to attain in the *post-operam* configuration seismic risk classes in the range B–A+ (i.e. %NBS index ≥ 60%).

The deduction is applied to a maximum expense obtained as €96,000 multiplied by the number of units in the building; it is spread by 5 annual equal shares. In 2019, the Budget Law (dicembre 2018) introduced a specific incentive for interventions on existing buildings aimed at simultaneously improving the energy efficiency (the so called "Ecobonus") and the seismic risk (tax deduction of 75–85% for interventions determining the increase of one or two classes, respectively).

Recently, Article 119 of the Relaunch Decree issued in 2020 (Decreto Legge 19 maggio 2020) introduced a further tax deduction incentive for energy efficiency seismic strengthening intervention increasing the tax deduction threshold to 110%.

4.6 Conclusions

The paper deals with the analysis of damage assessment of buildings and direct and indirect costs for reconstruction of about 11,500 residential building outside and inside historical centers (OHC and IHC) after the 2009 L'Aquila devastating earthquake. The goal of the work is to provide useful data for researchers involved in the definition of post-earthquake loss scenarios, for decision makers to establish mitigation policies and priorities in the aftermath of future earthquakes, and to insurance companies to value sound insurance premium for existing buildings in seismic prone areas.

The main outcomes can be summarized as follows:

- The mean unit repair cost as a function of usability rating which accounts for damage to several buildings components resulted €195.66/m^2 for lightly damaged buildings, and increased to €328.99/m^2 for building with medium damage up to €497.87/m^2 for severely damaged buildings;
- repair costs have been used to define values of %c_j (percentage of the reconstruction cost of new building equal to €1,350/m^2 at national level) associated to different empirical global Damage States, DS. The increasing trend of %c_j as a function of DS resulted: DS1, %c_1 = 2%; DS2, %c_2 = 10%; %c_3 = DS3 30%; %c_4 = D4 60%; %c_5 = DS5 100%;
- the repair cost of partitions and infills ranges from 43% to 58% of *Building Repair Costs, BRC*. This percentage rise to 81%–89% by adding the repair costs of construction system, plumbing and electrical systems as well as of windows and doors and enclosure systems which are commonly incorporated in hollow clay brick partitions and infills in the Mediterranean area;
- the repair of drift-sensitive components ranges from 63%–70% of *BRC* while 15%–21% relate to the repair of the acceleration-sensitive ones (i.e. roofs and chimneys, sanitary and other equipment, floor finishes).
- the mean unit strengthening intervention cost resulted equal to €46.29/m^2 for local strengthening intervention on lightly damaged buildings; €139.86/m^2; for local strengthening intervention on buildings with medium damage level; and €313.72/m^2 for global strengthening interventions on severely damaged buildings;
- the mean unit costs for strengthening intervention to attain 1% increase of the building safety index (i.e. +1%NBS) resulted equal to €7.8/m^2 and €10.7/m^2 for RC and masonry buildings, respectively, with severe structural damage;
- the mean unit costs for energy efficiency upgrade interventions resulted €38.89/m^2 or €68.93/m^2 for building with medium or severe damage;
- the accommodation costs for people assistance are a strong ratio of costs to be accounted for in the reconstruction process;
- the unit repair and strengthening costs of buildings Inside Historical Centres (IHC) are significantly higher than those related of buildings Outside Historical Centres (OHC) due to the extra costs needed to preserve, restore or repair valuable elements.

The data presented herein have been used in Italy to define the technical document specifically developed for the seismic risk classification of existing buildings and for the definition of the technical principles for exploiting tax deductions with respect to seismic strengthening interventions on private residential buildings (the so-called "*Sismabonus*").

Acknowledgements This study was performed within the framework of the PE 2019–2021 joint program DPC-ReLUIS, WP7: "Post-earthquake analysis: usability, damage, seismic design of repair and strengthening interventions"

References

Annex to OPCM no. 3779—"Guidelines for the implementation of the measures relative to the ordinance of the president of the council of ministers no. 3779 June 6, 2009". (In Italian), "Indirizzi per l'esecuzione degli interventi di cui all'Ordinanza del Presidente del Consiglio dei Ministri n. 3779 del 6 giugno 2009

Annex to OPCM no. 3790—"Guidelines for the implementation of the measures relative to the ordinance of the president of the council of ministers no. 3790 July 9, 2009". (In Italian), "Indirizzi per l'esecuzione degli interventi di cui all'Ordinanza del Presidente del Consiglio dei Ministri n. 3790 del 9 luglio 2009"

ATC (Applied Technology Council) (2005) ATC-20-1. Field manual: postearthquake safety evaluation of buildings, 2nd edn. Applied Technology Council, Redwood City, CA, USA

Baggio C, Bernardini A, Colozza R, Coppari S, Corazza L, Della Bella M, Di Pasquale G, Dolce M, Goretti A, Martinelli A, Orsini G, Papa F, Zuccaro G (2007) Field manual for post-earthquake damage and safety assessment and short term countermeasures (Pinto A, Taucer F eds), Translation from Italian: Goretti A, Rota M, JRC Scientific and Technical Reports, EUR 22868 EN-2007

Civil Protection Department (DPC) and Laboratories University Network of Seismic Engineering (ReLUIS) (2011) Guidelines for of structural elements, infill, and partitions (M. Dolce and M. G. Doppiavoce, eds.), ISBN 978-88-89972-29-8 (in Italian)

Cosenza E, Del Vecchio C, Di Ludovico M, Dolce M, Moroni C, Prota A, Renzi E (2018) The Italian guidelines for seismic risk classification of constructions: Technical principles and validation. Bull Earthq Eng 16:5905–5935

Decree of the head of the USRA January 21 (2013) n. 1 "Disciplina per la progettazione e la realizzazione degli interventi sugli edificI private, ubicate nei centri storici del comune di L'Aquila danneggiati dal sisma del 2009" Manual in Italian, https://usra.it/wp-content/uploads/2019/05/Manuale-istruzioni-scheda-progetto-parte-prima-aggiornato-al-Decreto-n.4.pdf

Decree of the head of the USRC February 06 (2014) n. 1 "Disposizioni per riconoscimento del contributo per gli interventi sull'edilizia privata nei centri storici dei Comuni del Cratere". IMC Model, Manual in Italian, http://mic.usrc.it/

Decreto Legge 19 maggio (2020) no. 34 Misure urgenti in materia di salute, sostegno al lavoro e all'economia, nonché' di politiche sociali connesse all'emergenza epidemiologica da COVID-19. (20G00052) (GU Serie Generale n.128 del 19-05-2020—Suppl. Ordinario n. 21) Decreto-Legge convertito con modificazioni dalla L. 17 luglio 2020, n. 77 (in S.O. n. 25, relativo alla G.U. 18/07/2020, n. 180)

Del Gaudio C, De Martino G, Di Ludovico M, Ricci P, Verderame GM (2017) Empirical fragility curves from damage data on RC buildings after the 2009 L'Aquila earthquake. Bull Earthquake Eng (2017)15:1425. https://doi.org/10.1007/s10518-016-0026-1

Del Vecchio C, Di Ludovico M, Prota A (2020) Repair costs of reinforced concrete building components: from actual data analysis to calibrated consequence functions. Earthq Spect 36(1):353–377

Di Ludovico M, Prota A, Moroni C, Manfredi G, Dolce M (2017a) Reconstruction process of damaged residential buildings outside historical centres after the L'Aquila earthquake: part I—"light damage" reconstruction. Bull Earth Eng 15(2):667–692

Di Ludovico M, Prota A, Moroni C, Manfredi G, Dolce M (2017b) Reconstruction process of damaged residential buildings outside historical centres after the L'Aquila earthquake—part II: "reconstruction". Bull Earth Eng 15(2):693–729

Dolce M (2010) Emergency and Post-Emergency Management of the Abruzzi Earthquake, Theme-leader lecture. In: 14th European conference on earthquake engineering, Ohrid, Macedonia, 3-8 September 2010, published in M. Garevski, A. Ansal (Eds.) in Earthquake Engineering in Europe, Springer, https://doi.org/10.1007/978-90-481-9544-2_1

Dolce M, Di Bucci D (2017) Comparing recent Italian earthquakes. Bull Earthq Eng 15:497–533

Dolce M, Speranza E, Giordano F, Borzi B, Bocchi F, Conte C, Di Meo A, Faravelli M, Pascale V (2019) Observed damage database of past Italian earthquakes: the Da. DO WebGIS. Bollettino di Geofisica Teorica ed Applicata 60(2)

Dolce M, Speranza E, Giordano F, Conte C, De Martino G (2019) Implementation of the national seismic risk prevention plan: the retrofit of strategic and relevant buildings (in Italian). XVIII ANIDIS Conference "Earthquake Engineering in Italya", Ascoli Piceno

D.P.C.M. February 4 (2013) no. 54—Definizione delle procedure per il riconoscimento dei contributi per la ricostruzione privata, conseguente agli eventi sismici del 6 aprile, adottato ai sensi dell'articolo 67-quarter, comma 9, del decreto legge 22 giugno 2012, n. 83, convertito, con modificazioni, dalla legge 7 agosto 2012, n. 134

Fico R, Gualtieri R, Pecci D, Mannella A, Di Ludovico M, Prota A (2017) Reconstruction model of residential buildings in the historical centres of the crater municipalities after L'Aquila 2009 earthquake. 16th World Conference on Earthquake Engineering, 16th WCEE 2017, Santiago Chile, January 9th to 12th 2017

Goretti A, Inukai M (2002) Post-earthquake usability and damage evaluation of reinforced concrete buildings designed not according to modern seismic codes. JSPS Short Term Fellowship, Final report, Servizio Sismico Nazionale, Dipartimento di Protezione Civile, Roma, Italy

Grünthal G (Ed.) (1998) European Macroseismic Scale 1998. European seismological commission, subcommission on engineering seismology, working group macroseismic scales, Cahiers du Centre Européen de Géodynamique et de Séismologie, 15, pp. 99, Luxemburg

Legge 30 dicembre 2018, no. 145 (Budget Law 2018) —GU Serie Generale n. 302 del 31-12-2018—Suppl. Ordinario n. 62)

Mannella A, Di Ludovico M, Sabino A, Prota A, Dolce M, Manfredi G (2017) Analysis of the and returning home in the reconstruction process of the 2009 L'Aquila earthquake. Sustainability 9(8):1395. https://doi.org/10.3390/su9081395

Ministry Decree no. 58 28/02/2017 Allegato A: linee guida per la classificazione del rischio sismico delle costruzioni (in Italian). Italian Ministry of Infrastructures and Trasport, Italy. http://www.mit.gov.it/normativa/decreto-ministeriale-numero-58-del-28022017

National Civil Protection Department, National risk assessment. Overview of the potential major disasters in Italy: seismic, updated December (2018)

NZSEE (New Zealand Society for Earthquake Engineering) (2009) Building safety evaluation during a state of emergency guidelines for Territorial Authorities. http://www.dbh.govt.nz/UserFiles/File/Building/information%20for/Building-Safety-Evaluation-during-State-of-Emergency.pdf

Ordinance of the President of the Council of Ministers, OPCM no. 3779, June 6 (2009) Urgent interventions to deal with seismic events occurring in the Abruzzo region on April 6, 2009 and other urgent civil protection provisions. (In Italian), OPCM n. 3779 del 6 giugno 2009—Ulteriori interventi urgenti diretti a fronteggiare gli eventi sismici verificatisi nella regione Abruzzo il giorno 6 aprile 2009 e altre disposizioni urgenti di protezione civile, Pubblicata nella Gazzetta Ufficiale n. 132 del 10 giugno 2009

Ordinance of the President of the Council of Ministers, OPCM no. 3790, July 9 (2009)—Urgent interventions to deal with seismic events occurring in the Abruzzo region on April 6, 2009 and other urgent civil protection provisions (In Italian), (In Italian), OPCM n. 3790 del 9 luglio—Ulteriori interventi urgenti diretti a fronteggiare gli eventi sismici verificatisi nella regione Abruzzo il giorno 6 aprile 2009 e altre disposizioni urgenti di protezione civile, Pubblicata nella Gazzetta Ufficiale n. 166 del 20 luglio 2009

Ordinance of the President of the Council of Ministers, OPCM no. 3881, June 11 (2010) Urgent interventions to deal with seismic events occurring in the Abruzzo region on April 6, 2009 and other urgent civil protection provisions (In Italian), OPCM n. 3881 del 11 giugno 2010—Ulteriori interventi urgenti diretti a fronteggiare gli eventi sismici verificatisi nella regione Abruzzo il giorno 6 aprile 2009 e altre disposizioni urgenti di protezione civile, Pubblicata nella Gazzetta Ufficiale n. 166 del 20 luglio 2009, Pubblicata nella Gazzetta Ufficiale n.152 del 2 luglio 2010

Presidency of Council of Ministers, Civil Protection Department. Final Note by the Deputy Commissioner for the Earthquake Emergency—January 29 2010. [http://www.protezionecivile.gov.it/res ources/cms/documents/PassaggioConsegne29_01_2010.pdf. Accessed 31 Jan 2017]

Severino M, Di Pasquale G (2002) Procedures for the post-earthquake reconstruction: analysis and proposals. Alinea. Procedure per la ricostruzione post-sisma: analisi e proposte (in Italian)

Chapter 5
The Modified Post-earthquake Damage Assessment Methodology for TCIP (TCIP-DAM-2020)

A. Ilki, O. F. Halici, M. Comert, and C. Demir

Abstract Post-Earthquake damage assessment has always been one of the major challenges that both engineers and authorities face after disastrous earthquakes all around the world. Considering the number of buildings in need of inspection and the insufficient number of qualified inspectors, the availability of a thorough, quantitative and rapidly applicable damage assessment methodology is vitally important after such events. At the beginning of the new millennia, an assessment system satisfying these needs was developed for the Turkish Catastrophe Insurance Pool (TCIP, known as DASK in Turkey) to evaluate the damages in reinforced concrete (RC) and masonry structures. Since its enforcement, this assessment method has been successfully used after several earthquakes that took place in Turkey, such as 2011 Van Earthquake, 2011 Kutahya Earthquake, 2019 Istanbul Earthquake and 2020 Elazig Earthquake to decide the future of damaged structures to be either 'repaired' or 'demolished'. Throughout the years, the number of research activities focusing on the reparability of earthquake-damaged structures has increased, which is a purposeful parameter in the determination of buildings' future after earthquakes. Accordingly, TCIP initiated a research project with a sole aim to regulate and reevaluate the damage assessment algorithm based on the results of state-of-the-art scientific research. This chapter presents the new version of the damage assessment methodology for reinforced concrete structures which was developed for TCIP (TCIP-DAM-2020). In addition, an application of the developed damage assessment algorithm on an earthquake-damaged reinforced concrete building which was struck by Kocaeli (1999) earthquake is presented.

A. Ilki (✉) · C. Demir
Istanbul Technical University, Civil Engineering Faculty, Maslak, Istanbul 34469, Sariyer, Turkey
e-mail: ailki@itu.edu.tr

O. F. Halici
Department of Civil Engineering, MEF University, 34396 Maslak, Istanbul, Sariyer, Turkey

M. Comert
RISE Engineering, 34398 Maslak, Istanbul, Sariyer, Turkey

S. Akkar et al. (eds.), *Advances in Assessment and Modeling of Earthquake Loss*,
Springer Tracts in Civil Engineering,
https://doi.org/10.1007/978-3-030-68813-4_5

5.1 Introduction

Since the second quarter of the twentieth century, a number of destructive earthquakes that took place all around the world caused total or partial collapse of structures and resulted in a great number of casualties and negative economic impacts (i.e. 1940 El Centro, 1967 Mudurnu, 1985 Mexico City, 1995 Kobe, 1999 Kocaeli, 2009 L'Aquila, 2011 Christchurch, 2011 Tohoku and 2017 Puebla earthquakes). Post-earthquake site investigations after damaging seismic events revealed that the number of structures demanding a damage inspection could be extraordinary (AIJ/JSCE/JGS 2001; Alberto et al. 2018; Alexander 2010; Erdik 2000; Kazama and Noda 2012; Marquis et al. 2017) and the insufficient number of qualified inspectors makes the execution of damage assessment a great challenge to accomplish on the way of returning back to everyday life. After disastrous earthquakes, a consistent damage assessment methodology is needed for re-establishing the evacuated structures for the accommodation of the locals and prohibiting the residents to enter the critically damaged structures that might collapse during probable aftershocks. Implementing a reliable methodology is vital for avoiding the unnecessary demolition of damaged structures which creates additional burdens to individuals and national economies. Furthermore, considering the large number of buildings in need of inspection after damaging earthquakes, the assessment methodology needs to be rapidly applicable and straight forward. Past damage assessment experiences gained after a number of earthquakes that took place in Turkey (i.e. 1995 Dinar Earthquake, 1998 Adana Earthquake, 1999 Kocaeli Earthquake and 1999 Duzce Earthquake) also indicated that since the assessors on-site have different backgrounds and experience levels, an objective damage assessment and decision-making is not possible without a quantitative and systematic damage assessment algorithm.

In the year 1999, after the earthquakes that struck the north-western part of Turkey (AIJ/JSCE/JGS 2001; Aydan et al. 2000), the Turkish government implemented a change on the state aid policy to the earthquake victims whose houses are collapsed or damaged during seismic events. The new regulation stated compulsory seismic insurance of structures. Consequently, Turkish Catastrophe Insurance Pool (TCIP) was established in 2000 to execute the compulsory earthquake insurance. In 2002, TCIP appointed researchers to develop a consistent, rapid and easy-to-apply damage assessment method to be benefitted after earthquakes. Accordingly, a methodology satisfying the fundamental characteristics expected from a reliable damage assessment algorithm mentioned above was developed for TCIP (TCIP-DAM-2002) for the two most common structural systems used in Turkey; Reinforced Concrete (RC) and masonry structures, respectively (Boduroglu et al. 2013; Ilki et al. 2013). During its development, the methodologies used in widely accepted guidelines that assert the recommended practices for the post-earthquake damage assessment have been benefitted (Anagnostopoulos et al. 2004; Baggio et al. 2007; Grünthal 1998; FEMA 306 1998; New Zealand Society for Earthquake Engineering (NZSEE) 2009; Japan and building disaster prevention association (JBDPA) 2015). The developed damage assessment algorithm, similar to the methodology used in Japan (Japan and building disaster prevention association (JBDPA) 2015), determines the building safety based on the residual energy dissipation capacity of structural members that degrades due to seismic actions. Since its development, a number of adjustments and improvements have been implemented to make the method easier to apply. For instance, in 2015, a quick inspection algorithm was implemented for those structures having a plan area less than 400 m^2, and whose number of stories above the ground level or rigid basement is less than eight. The main motivation for quick assessment methodology was to complete the damage assessment more rapidly for regular structures built in Turkey. The developed damage assessment system, which includes both detailed and quick inspection algorithms, has been presented in a number of education seminars carried out around Turkey (e.g., Istanbul, Ankara, Izmir, Canakkale, Kocaeli, Elazig, Manisa, etc.) that were organized by different institutions (i.e., Ministry of Environment and Urban Planning, TCIP and Turkish Chamber of Civil Engineers). This method has been successfully used by TCIP in the decision-making processes of earthquake-damaged structures after a number of earthquakes that took place in Turkey, including Van (2011), Kutahya (2011), Istanbul (2019), Elazig (2020) earthquakes. The experience and site observations gained through the application of the TCIP-DAM-2002 revealed that, there is a need for an even quicker methodology.

In recent years, TCIP has established a new action to advance the damage assessment algorithm. One incentive for this action was to make the damage assessment framework, if possible, even quicker and easier to apply without any compromise in reliability and objectivity. Another encouragement was to consider the state-of-the-art scientific research executed in the last two decades which can be benefitted in the further development of the damage assessment system either by modifying or further validating the theory behind the methodology. Also, in the last two decades, the number of code-complying structures, which are designed and detailed in accordance with the capacity design principle, is considerably increased. In addition to the damages in vertical members, these structures are expected to exhibit damages in horizontal members. Hence, a need has arisen for a damage assessment method that takes into account the damages formed on the beams as well. Apart from that, instead of an assessment algorithm that mechanically determines the limits to repair or demolish the earthquake-damaged buildings, a novel approach that estimates and considers the economic feasibility of the repair applications in post-earthquake decision-makings would be more beneficial (Ludovico et al. 2017a, b; Martino et al. 2017). By doing so, the algorithm should also consider the cost of nonstructural members' repair because of the fact that a great portion of the budget reserved for the repair applications of earthquake-damaged structures is spent on the non-structural members (Cardone and Perrone 2017; Taghavi and Miranda 2003; Vecchio et al. 2018, 2020).

A novel damage assessment methodology in accordance with the needs stated above has been developed. More than 100 experimental test results obtained from literature were benefitted in the determination of member damage limits and damage modification factors used in the methodology. Also, more than 200 structural performance analyses and 80,000 cost analyses with different damage scenarios have been carried out for the determination of the limits for building damage categories. In this manuscript, although the method is applicable to both masonry and RC structures, due to page limitations, only the damage assessment algorithm developed for RC structures is presented.

5.2 The Revised Version of TCIP Damage Assessment System

5.2.1 Building Damage Categories

After earthquakes, structures in seismic zones suffer different levels of damage as a result of a process in which various parameters play a role, including structural system characteristics, design and construction errors, ground motion characteristics and soil conditions, etc. During post-earthquake damage inspections, the structural damages caused by the earthquake effects shall be observed in the form of cracking, crushing or spalling of concrete, rupture or buckling of reinforcements, sagging in the horizontal structural members, residual drifts, uniform or differential settlements and tilting of the building, etc. The revised version of the damage assessment system uses site observations and simple measurements as input. However, in the background, the building damage categorization is determined by the evaluation of mechanical and financial feasibility criteria based on the input data. The revised system defines six building damage categories as follow.

5.2.1.1 Undamaged Building

This damage category corresponds to a condition where there is no earthquake damage in vertical (i.e. columns and shear walls) or horizontal (i.e. beams) load-bearing structural members. However, it needs to be emphasized that, the structure might contain some damages formed before the earthquake action typically due to time and environmental effects (e.g. corrosion, shrinkage, freeze–thaw cracks) or other mechanical effects except earthquake (e.g. excessive vertical load, soil settlement). The building maintains its pre-earthquake performance and capacity.

5.2.1.2 Slightly Damaged Building

In the case of slightly damaged building, the vertical and horizontal members that form the structural system of the building suffered limited damages in such a way that the damaged members either do not entail any repair or require relatively simple repair applications. Nonstructural elements such as infill walls might experience some damages, but, in general, they are easily repairable. The building predominantly preserves its pre-earthquake performance and capacity.

5.2.1.3   Moderately Damaged Building

Due to the damages in the vertical and horizontal structural members, the performance and capacity of the structure can be decreased to a certain degree in comparison to that of pre-earthquake condition. In addition to the damages in the structural elements, extensive damages in nonstructural elements can be observed. Still, with further investigations and comprehensive engineering evaluations, it is technically and economically possible to repair and strengthen the building.

5.2.1.4   Heavily Damaged Building

In heavily damaged buildings, the damages in the structural members can reach to severe levels. In addition, many of the nonstructural members of the building are substantially damaged. The building may have lost a significant amount of its pre-earthquake performance and capacity. Due to the necessity of wide-scale and comprehensive structural interventions, the repair and strengthening applications for the structure may be far from being economically feasible. Therefore, demolition-and-reconstruction is generally a more convenient option for these buildings.

5.2.1.5   Building to be Urgently Demolished

The buildings where a partial collapse has occurred in at least one story, or the buildings exhibiting easily observable residual displacements are classified in this category. The existing condition of these buildings poses danger to the safety of life and property. Hence, the demolition of these buildings should be prioritized.

5.2.1.6   Collapsed Building

The structural system lost its integrity and the building is collapsed partially or completely. The vertical and horizontal load carrying capacity of the building is entirely eliminated.

5.2.2   Damage Categories for RC Members

The damage categorizations of vertical and horizontal RC structural members are made in accordance with the rules and limits defined in this section and the observed damages. There are five member damage categories defined to be used in the damage assessment algorithm. Details of the damage categories, whose limits are presented in Table 5.1, are given in Sect. 5.2.2.1–5.2.2.5.

5.2.2.1 Type O Damage Category

Regardless of the damages caused by environmental and time-dependent effects (e.g. corrosion, creep, shrinkage and non-seismic ground settlements), vertical and horizontal structural members which do not contain any damage caused by earthquake effects are assigned to Type O damage category.

5.2.2.2 Type A Damage Category

Vertical and horizontal RC structural members which contain at least one crack with a maximum residual width of 0.5 mm that was formed due to earthquake actions are defined as Type A damaged element. The categorization is carried out regardless of whether the cracks are formed due to bending or shear effects. Typical examples of Type A damages are presented in Fig. 5.1 for bending and shear cracks.

5.2.2.3 Type B Damage Category

Those vertical and horizontal RC structural members that contain at least one crack between 0.5 and 3 mm in width or exhibit slight concrete crushing limited to cover are categorized as Type B damaged structural elements. Figure 5.2 shows typical bending and shear damages that are considered to be Type B damage.

5.2.2.4 Type C Damage Category

Vertical and horizontal structural members containing at least one earthquake-induced crack whose width is more than 3 mm or exhibit concrete cover spalling are categorized as Type C damaged structural members. The structural elements

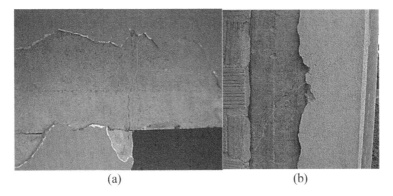

(a)   (b)

Fig. 5.1 Examples of Type A Damage Category; **a** flexural damage; **b** shear damage

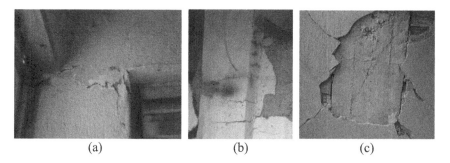

(a)   (b)   (c)

Fig. 5.2 Examples of Type B Damage Category; **a** concrete crushing; **b** flexural crack; **c** shear damage

showing negligible buckling of reinforcement that do not significantly deviate from its alignment are also considered in this damage category. In the previous version of the damage assessment methodology, the members with buckled reinforcement were assigned to Type D damage category. However, as will be discussed in Sect. 5.2.3.2, the structures with Type D vertical elements will be directly assigned to Heavily Damaged building category. Hence, in order not to categorize a whole structure as Heavily Damaged because of a single vertical member with an indistinct reinforcement buckling, this damage level is included in Type C damage category. Figure 5.3 presents representative structural members that are deemed to be categorized as Type C damage.

5.2.2.5 Type D Damage Category

Vertical and horizontal structural members exhibiting core concrete crushing, reinforcement buckling, stirrup rupture or distinctive residual deformations that are formed due to earthquake actions are categorized as Type D damaged elements. Descriptive structural elements having Type D damage are presented in Fig. 5.4.

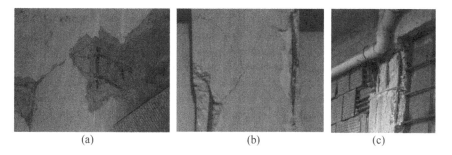

(a)   (b)   (c)

Fig. 5.3 Examples of Type C Damage Category; **a** flexural damage; **b** shear damage; **c** slight buckling of reinforcement

(a)   (b)   (c)

Fig. 5.4 Examples of Type D Damage Category; **a** flexural damage; **b** shear damage; **c** buckling of reinforcement and core crushing

5.2.3 Damage Assessment Algorithm

The damage assessment algorithm consists of a two-stage procedure; (i) exterior assessment and (ii) interior assessment. The evaluation begins with the exterior assessment. Depending on the damage condition of the building, the inspectors proceed to the interior assessment stage with one of the methods defined in Sect. 5.2.3.2.

5.2.3.1 Exterior Assessment

In this stage, visual inspections and measurements will be carried out in regard to the general condition of the subject structure. The exterior assessment will be completed with respect to the following inspection processes.

- If the building is entirely collapsed, the damage categorization of the building is determined as Collapsed Building. If a partial collapse is observed (Fig. 5.5a), the damage assessment is concluded by classifying the structure as Building to be Urgently Demolished.
- If the permanent horizontal residual displacement measured at any story in the building is greater than 1% of the corresponding story height, the building is categorized as Heavily Damaged Building and the assessment is finished. If the horizontal residual displacement at any story is greater than 3% of the corresponding story height, the building is classified as Building to be Urgently Demolished. Figure 5.5b shows a building that suffered from excessive residual displacements that occurred due to seismic actions. The story height (h) and the horizontal residual displacement (d) are schematically illustrated in Fig. 5.6b.
- If the structure exhibits a rigid rotation greater than 2° due to different settlements caused by earthquake effects, the damage category of the building is defined as Heavily Damaged Building and the assessment is terminated. If the rigid rotation

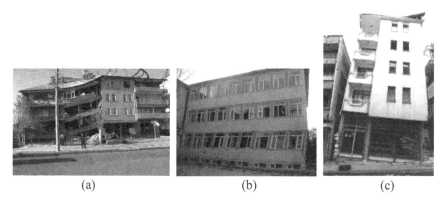

Fig. 5.5 Damaged structures; partial collapse; **b** excessive residual drift; **c** tilting

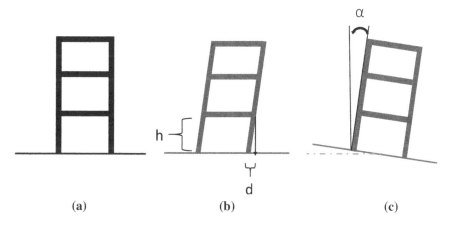

Fig. 5.6 **a** undamaged building; **b** building with residual drift; **c** tilted building

is greater than 4°, the damage assessment is concluded by categorizing the structure as Building to be Urgently Demolished. Figure 5.5c shows a structure that significantly tilted due to rigid rotation at the base. A schematic illustration of the tilting angle is presented in Fig. 5.6c.

If none of the damage conditions stated above exist in the subject structure and there are no obvious structural and nonstructural damages making the entry of the building dangerous, the inspectors proceed to interior assessment stage. Before entering the building, the inspector crew should bear in mind the possibility of aftershock occurrence and need to follow the safety measures.

Table 5.1 Damage limits for RC members

Damage category	Residual crack width	Compression damage
Type O	–	–
Type A	≤0.5 mm	–
Type B	0.5 mm < w ≤ 3 mm	Cover crushing
Type C	>3 mm	Cover spalling
Type D	–	Buckling of reinforcement, core crushing and residual displacement

5.2.3.2 Interior Assessment

In the interior assessment stage, the building damage category is determined based on the damage categories of the vertical and horizontal structural members of the structure that are categorized according to Table 5.1. During the inspection, if the subject structure contains at least one vertical structural member that is categorized as Type D, the building is categorized as Heavily Damaged. In addition, based on the damage conditions given in rapid and detailed inspection procedure, the structures can be classified as Building to be Urgently Demolished. On the other hand, if all the structural members are undamaged, the building is categorized as Undamaged. If there is no vertical structural element categorized as Type D and not all of the structural members are undamaged in the inspected building, the building damage category is determined by applying one of the interior assessment procedures (i.e. rapid inspection procedure and detailed inspection procedure) at the most severely earthquake-damaged story of the structure. It needs to be emphasized that, the rapid inspection procedure is developed and designed to be suitable for the majority of building type RC structures. On site, the inspectors are mostly expected to use the rapid inspection procedure. For the exceptional cases where the building is not in the application limits of the rapid inspection procedure, the inspectors will apply the detailed procedure. In interior assessment procedures, the limit for earthquake-damaged structures to be categorized as Slightly Damaged is determined based on the loss in the structural performance that is caused by earthquake damages. For this, more than 200 seismic performance analyses have been executed considering different damage case scenarios. On the other hand, the limit for Heavily Damaged structures is determined from the repair cost of structural and nonstructural members. The repair costs of structural members exhibiting earthquake damages given in Table 5.1 are obtained from market investigations. In accordance with the findings obtained in the existing researches (Cardone and Perrone 2017; Taghavi and Miranda 2003; Vecchio et al. 2018, 2020) the repair and the cosmetic cost of nonstructural members (i.e., infill walls, floor finishes, ceiling floors, etc.) is approximately assumed to be twice of the structural members' repair cost. In the cost analyses, by relating the structural damages with the structural and nonstructural repair costs, the damage level where the cost of repair becomes financially infeasible is defined to be the limit for the structures to be categorized as Heavily Damaged. This limit is obtained from the results of more than 80,000 cost analyses representing different damage case scenarios.

Rapid Inspection Procedure

Rapid inspection method can be employed for the damage assessment of structures whose Plan Area (PA) is less than 600 m² and the number of stories above the ground level or rigid basement is less than or equal to 10. In this procedure, the building damage category is obtained based on the number of damaged structural members which are categorized according to the member damage categories defined in Sect. 5.2.2. The number limits for vertical and horizontal structural elements with certain damage categories are generated based on the PA of the inspected structure. Building damage category of inspected structure is determined by considering the following damage limits.

- Damage limits for vertical structural members:
  - The case where the number of vertical members categorized as Type B is less than PA/100 and there is no vertical member classified as Type C and Type D
  - The case where the number of vertical members categorized as Type B is greater than or equal to PA/100 or the number of vertical members categorized as Type C is at least one but smaller than PA/200 and no vertical member is classified as Type D
  - The case where the number of vertical members categorized as Type C is greater than or equal to PA/200 but smaller than PA/75 and no vertical member is classified as Type D
  - The case where the number of vertical members categorized as Type C is greater than or equal to PA/75 or there is at least 1 vertical member classified as Type D.

- Damage limits for horizontal structural members:
  - The case where no horizontal members are categorized as either Type C or Type D
  - The case where the number of horizontal structural members categorized as Type C and Type D is at least one but less than PA/50
  - The case where the number of horizontal members categorized as Type C and Type D is greater than or equal to PA/50 but less than PA/20
  - The case where the number of horizontal members categorized as Type C and Type D is greater than or equal to PA/20.

The damage assessment is concluded through the determination of relevant damage ranges outlined in Table 5.2 for both vertical and horizontal structural

Table 5.2 Damage limits for rapid damage assessment methodology

Determination of Building Damage Category		Vertical Structural Members			
		B < PA/100 and C+D = 0	B ≥ PA/100 or 1≤ C < PA/200 and D = 0	PA/200≤ C <PA/75 and D = 0	C ≥ PA/75 or D ≥ 1
Horizontal Structural Members	C+D = 0	SLIGHTLY DAMAGED	MODERATELY DAMAGED	MODERATELY DAMAGED	HEAVILY DAMAGED
	1≤ C+D <PA/50	MODERATELY DAMAGED	MODERATELY DAMAGED	HEAVILY DAMAGED	HEAVILY DAMAGED
	PA/50≤ C+D < PA/20	MODERATELY DAMAGED	HEAVILY DAMAGED	HEAVILY DAMAGED	HEAVILY DAMAGED
	C+D ≥ PA/20	HEAVILY DAMAGED	HEAVILY DAMAGED	HEAVILY DAMAGED	HEAVILY DAMAGED

members and the building damage category is obtained by the intersection of these intervals. The rapid damage assessment algorithm is presented in Fig. 5.7.

Detailed Inspection Procedure

If the subject structure is not suitable for the rapid inspection, the detailed procedure can be applied regardless of limits for the base area or the number of stories of the building. In this examination, Weighted Damage Percentage for Vertical Members (WDPVM) and the number of damaged horizontal members are determined at the inspected story based on the observed damage categories and damage modifiers presented in Table 5.3. More than 100 experimental test results have been exploited in the determination of these factors. For the corresponding damage conditions stated in Table 5.1, the dissipated energies were compared with the total energy dissipation capacity of the specimens. The factors in Table 5.3 represents the ratio of the dissipated energy to the total energy dissipation capacity of the structural members.

The damage level of each vertical structural member is weighted with its cross-sectional area. Accordingly, the calculation of WDPVM is carried out with respect to Eq. (5.1) where O, A, B and C stand for the total cross-sectional area of the vertical members assigned to Type O, Type A, Type B and Type C damage categories, respectively. Since the case of observing at least one vertical member with Type D

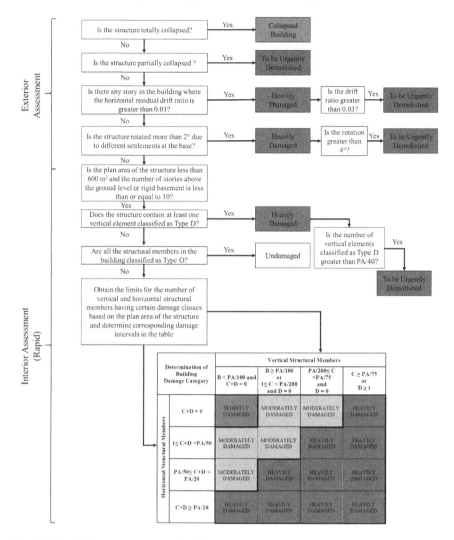

Fig. 5.7   Rapid damage assessment algorithm

Table 5.3 Damage modification factors (λ) for RC elements

Member damage category	Damage modification factor
O	0.00
A	0.20
B	0.40
C	0.70
D	1.00

damage category directly leads the damage assessment procedure to a conclusion where the damage category of the building is determined as either Heavily Damaged or Building to be Urgently Demolished, the vertical members with Type D damage category are excluded in Eq. (5.1). The coefficients (0.20, 0.40 and 0.70) are the damage modification factors (λ) for the corresponding damage categories given in Table 5.3.

$$WDPVM = \frac{A \times 0.20 + B \times 0.40 + C \times 0.70}{O + A + B + C} \times 100 \qquad (5.1)$$

A weighted damage percentage is not calculated for the horizontal structural members because of the practical concerns and the fact that the variations in the beam sizes are considerably low in comparison to that in the vertical members. It is sufficient to determine the number of Type C and Type D members by considering the damage limits defined in Sect. 5.2.2.

The damage category of the inspected building is determined based on the damage categories observed in vertical and horizontal members together with the damage percentage obtained from Eq. (5.1) and the limits that are determined based on the PA of the structure. For vertical and horizontal structural members, the following damage limits are defined for the detailed damage assessment algorithm.

- Damage limits for vertical structural members:

  - The case where WDPVM is less than 10 and no vertical member is categorized as either Type C or Type D
  - The case where WDPVM is greater than or equal to 10 but less than 20 or at least one vertical member is categorized as Type C and no vertical member is classified as Type D
  - The case where WDPVM is greater than or equal to 20 but less than 40 and no vertical member is classified as Type D
  - The case where WDOVM is greater than or equal to 40 or there is at least 1 vertical member classified as Type D.

- Damage limits for horizontal structural members:

  - The case where no horizontal members are categorized as either Type C or Type D
  - The case where the number of horizontal structural members categorized as Type C and Type D is at least one but less than PA/50
  - The case where the number of horizontal members categorized as Type C and Type D is greater than or equal to PA/50 but less than PA/20
  - The case where the number of horizontal members categorized as Type C and Type D is greater than or equal to PA/20.

Similar to the detailed inspection procedure, the building damage category for inspected buildings is obtained by determining the damage intervals for both vertical and horizontal members in accordance with Table 5.4 for both vertical and horizontal

Table 5.4 Damage limits for detailed damage assessment methodology

Determination of Building Damage Category		Vertical Structural Members			
		WDPVM< 10 and C+D = 0	C ≥ 1 or 10≤WDPVM<20 and D = 0	20≤WDPVM<40 and D = 0	WDPVM ≥ 40 or D ≥ 1
Horizontal Structural Members	C+D = 0	SLIGHTLY DAMAGED	MODERATELY DAMAGED	MODERATELY DAMAGED	HEAVILY DAMAGED
	1≤ C+D <PA/50	MODERATELY DAMAGED	MODERATELY DAMAGED	HEAVILY DAMAGED	HEAVILY DAMAGED
	PA/50≤ C+D <PA/20	MODERATELY DAMAGED	HEAVILY DAMAGED	HEAVILY DAMAGED	HEAVILY DAMAGED
	C+D ≥ PA/20	HEAVILY DAMAGED	HEAVILY DAMAGED	HEAVILY DAMAGED	HEAVILY DAMAGED

structural members. The building damage category is obtained by the intersection of these intervals. The detailed damage assessment algorithm for reinforced concrete structures is outlined in Fig. 5.8.

5.3 Case Study: Assessment of a Structure Damaged After 1999 Kocaeli Earthquake

An earthquake damaged structure investigated after the 1999 Kocaeli earthquake is re-evaluated according to TCIP-DAM-2020 method presented above. The details about the structure, location and observed damages are obtained from the earthquake report prepared by Architectural Institute of Japan in 2001 (AIJ/JSCE/JGS 2001). The building consisted of six stories and was made of reinforced concrete vertical and horizontal structural members. The building was located in Degirmendere district of Kocaeli and was under construction when the earthquake struck.

The structural system of the building did not exhibit partial or total collapse after the earthquake. However, many of the infill walls in the structure were heavily damaged. The walls on the cantilever beams were constructed with AAC blocks and the rest of the infills were built with hollow clay bricks. General views of the structure after the earthquake are shown in Fig. 5.9. The outer dimensions of the structure are

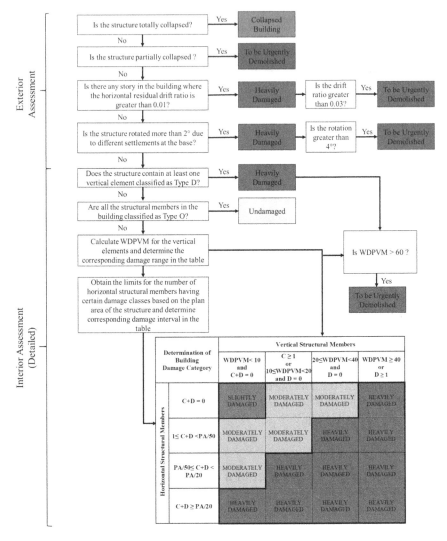

Fig. 5.8 Detailed damage assessment algorithm

11.6 and 11.4 m in X and Y directions, respectively. The PA of the structure is approximately 125 m². The plan view of the first story of the structure is shown in Fig. 5.10. The report (AIJ/JSCE/JGS 2001) stated that typical columns have 250 × 500 mm cross-section dimensions and have 8 longitudinal reinforcing bars with a diameter of 16 mm, which corresponds to 1.29% of a longitudinal reinforcement ratio. No information was given regarding stirrup diameter, spacing and hook details. Typical beams in the building have 200 × 500 mm cross-section dimensions. It was stated that bars with diameters of 12 and 14 mm were used as longitudinal reinforcements in the beams together with 6 mm stirrups with a spacing of 250 mm.

Table 5.5 Damage limits obtained for the case study structure

Determination of Building Damage Category		Vertical Structural Members			
		B < 1.25 and C+D = 0	B ≥ 1.25 or 1≤ C < 0.63 and D = 0	0.63 ≤ C < 1.67 and D = 0	C ≥ 1.67 or D ≥ 1
Horizontal Structural Members	C+D = 0	SLIGHTLY DAMAGED	MODERATELY DAMAGED	MODERATELY DAMAGED	HEAVILY DAMAGED
	1≤ C+D <2.5	MODERATELY DAMAGED	MODERATELY DAMAGED	HEAVILY DAMAGED	HEAVILY DAMAGED
	2.5≤ C+D < 6.25	MODERATELY DAMAGED	HEAVILY DAMAGED	HEAVILY DAMAGED	HEAVILY DAMAGED
	C+D ≥ 6.25	HEAVILY DAMAGED	HEAVILY DAMAGED	HEAVILY DAMAGED	HEAVILY DAMAGED

Fig. 5.9 Views of the building; **a** Western side; **b** Southwestern side (Modified from AIJ/JSCE/JGS 2001)

The most damaged story

The most damaged story

(a)   (b)

The plan area of the structure is smaller than 600 m^2 and the number of stories above the ground level is less than or equal to 10. Hence the damage assessment can be executed by following the rapid evaluation algorithm.

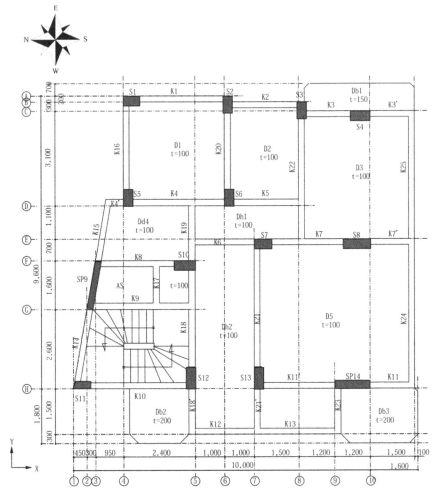

Fig. 5.10 Plan view of the evaluated building (AIJ/JSCE/JGS 2001) (mm)

Exterior Assessment

As shown in Fig. 5.7, the evaluation procedure starts with exterior assessment. The structure did not exhibit partial or total collapse and a residual drift that is greater than 1%. Also, no rigid rotation at the base that is greater than 2° was observed. The answer to these steps stated in the exterior assessment is 'No', hence the building damage category cannot be obtained as Heavily Damaged or Building to be Urgently Demolished from the exterior assessment phase. Thus, the damage assessment procedure continues with the interior assessment.

Interior Assessment

The interior assessment begins at the most damaged story in the building, which in this case is the first floor. After Kocaeli (1999) earthquake, the post-earthquake damage assessment of the structure was carried out according to the 1991 version of the Japanese damage assessment guideline (Japan and building disaster prevention association (JBDPA) 2015), which was in force at that time in Japan. The 2015 version of the guideline (Japan and building disaster prevention association (JBDPA) (2015) evaluates the post-earthquake condition of buildings considering total collapse mechanism where damages in both columns and beams are considered. Conversely, the 1991 version assumed soft-story collapse mechanism which only took into account the degradations in the shear strength of the vertical members.

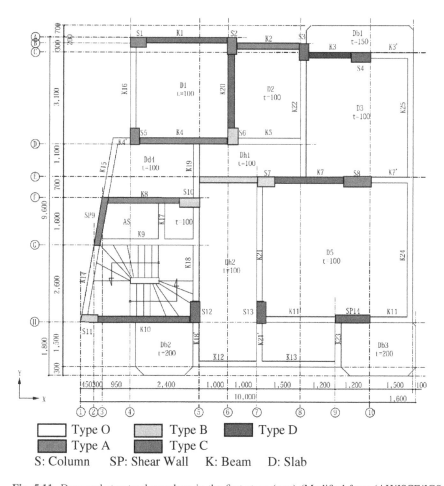

Fig. 5.11 Damaged structural members in the first story (mm) (Modified from (AIJ/JSCE/JGS 2001)

The earthquake report (AIJ/JSCE/JGS 2001) did not state the detailed damages for all elements (i.e. residual crack width, crushing of cover concrete, etc.), instead, the damage categories of vertical and horizontal structural members which were assigned in accordance with the 1991 Japanese method (Japan and building disaster prevention association (JBDPA) 2015) were given. By considering the member damage categories defined in the Japanese method (Japan and building disaster prevention association (JBDPA) 2015) and TCIP-DAM-2020 (Table 5.1), the reported member damages were converted to the corresponding member damage categories defined in TCIP-DAM-2020 method. The vertical and horizontal structural members with different damage categories are emphasized with different colors on the plan view of the first story shown in Fig. 5.11. The representative photos of the structural damages observed in the vertical and horizontal members are presented in Fig. 5.12.

No vertical structural members are categorized as Type D; therefore, the building cannot be directly categorized as Heavily Damaged. Also, because of the existence of damaged structural members, the structure cannot be directly categorized as Undamaged. Hence, the damage category is determined via the number limits defined for structural members with certain damage categories (Table 5.2). In the first story, five vertical members are categorized as Type C and the total number of horizontal elements categorized as Type C and Type D is five. Based on the PA of the structure (125 m^2) the limits for the number of damaged vertical and horizontal members are presented in Table 5.5. The number of vertical members with Type C damage category is greater than 1.67 and the number of horizontal members with Type C and Type D damage category is between 2.5 and 6.25. By intersecting the corresponding damage

(a) Cover spalling in SP14, Type C   (b) Cover spalling in S12, Type C

(c) Shear crack and cover spalling in K7, Type C

Fig. 5.12 Damaged members (Modified from (AIJ/JSCE/JGS 2001)

intervals determined for vertical and horizontal members, the damage assessment performed according to TCIP-DAM-2020 is concluded by categorizing the building as 'Heavily Damaged'. After the earthquake, the building was categorized as 'Very Heavy Damage' according to the 1991 version of the Japanese guideline (Japan and building disaster prevention association (JBDPA) 2015). Also, at that time, the building was evaluated based on EMS-98 (Grünthal 1998) as well. The damage grade of the structure by this code was ranked as Damage Grade 3 which corresponds to 'Substantial to Heavy Damage'.

5.4 Concluding Remarks

In this paper, the general framework of the new version of the damage assessment algorithm developed for TCIP is presented (TCIP-DAM-2020). In the development of rapid and detailed assessment algorithms, the observations gained through the execution of 200 numeric structural performance analyses and 80,000 repair cost analyses are considered. Also, the results of more than 100 experimental tests obtained from the literature have been benefitted to determine the modification factors that represent the behavior of earthquake-damaged structural members.

One of the most important improvements in the new algorithm is the enlargement of the application area of the rapid assessment method so that the method can be applied to the majority of the building stock in Turkey. This method enables the determination of the building damage category based on the number of damaged vertical structural members and the plan area of the inspected structure without calculating damage percentages for vertical members. If a damaged structure does not fulfill the geometric limitations defined in the rapid assessment method, the detailed method is implemented. Also, in the new version, both in rapid and detailed methods, if the building has at least one vertical member with Type D damage category, the building damage class can be easily determined as Heavily Damaged or Building to be Urgently Demolished. With TCIP-DAM-2020, the damage assessment applications are envisioned to be carried out in a much quicker way which would enable the inspectors to evaluate more earthquake-damaged structures in a certain time period. Considering the number of buildings in need of damage inspection and the lack of qualified personnel after earthquakes, this is deemed to be the most beneficial feature of TCIP-DAM-2020.

Another significant improvement in the new version of the damage assessment algorithm for RC structures is the inclusion of the damages that take place in the horizontal structural members in the determination of building damage category. This is especially beneficial for the post-earthquake damage assessment of code-complying structures whose portion in the building stock is constantly increasing due to the transformation of cities. All in all, the new damage assessment algorithm which is based on a broad scientific background and experimental and numerical analyses will enable quicker post-earthquake damage assessments without any compromise in objectivity and reliability.

Acknowledgements The effort of Turkish Catastrophe Insurance Pool to initiate a research project that forms the base of the work presented in this paper is thankfully acknowledged. The authors also would like to express their appreciation to the project consultants: Prof. Dr. Kutay Orakcal, Prof. Dr. Erdem Canbay and Prof. Dr. Bilge Doran for their contributions and constructive comments to the project.

References

AIJ/JSCE/JGS (2001) Report on damage investigation of the 1999 Kocaeli Earthquake in Turkey. Technical Report by joint reconnaissance team of architectural institute of Japan, Japan Society of Civil Engineers, The Japanese Geotechnical Society

Alberto Y, Otsubo M, Kyokawa H, Kiyota T, Towhata I (2018) Reconnaissance of the 2017 Puebla, Mexico earthquake. Soils Found 58(5):1073–1092

Alexander DE (2010) The L'Aquila earthquake of 6 April 2009 and Italian Government policy on disaster response. J Nat Resour Policy Res 2(4):325–342

Anagnostopoulos SA, Moretti M, Panoutsopoulou M, Panagiotopoulou D, Thoma T (2004) Post earthquake damage and usability assessment of buildings: further development and applications. Final report

Aydan Ö, Ulusay R, Kumsar H, Tuncay E (2000) Site investigation and engineering evaluation of the Düzce-Bolu earthquake of November 12 (1999) Technical Report, Turkish Earthquake Foundation. TDV/DR 95:51

Baggio C, Bernardini A, Colozza R, Corazza L, Della Bella M, Di Pasquale G, Dolce M, Goretti A, Martinelli A, Orsini G, Papa F (2007) Field manual for post-earthquake damage and safety assessment and short term countermeasures (AeDES). European Commission—Joint Research Centre—Institute for the Protection and Security of the Citizen, EUR, 22868

Boduroglu H, Ozdemir P, Binbir E, Ilki A (2013) Seismic damage assessment methodology developed for Turkish compulsory insurance system. In: Proceedings of the 9th annual international conference of the international institute for infrastructure renewal and reconstruction. Brisbane, Australia

Cardone D, Perrone G (2017) Damage and loss assessment of pre-70 RC frame buildings with FEMA P-58. J Earthq Eng 21(1):23–61

Del Vecchio C, Di Ludovico M, Pampanin S, Prota A (2018) Repair costs of existing RC buildings damaged by the L'Aquila earthquake and comparison with FEMA P-58 Predictions. Earthq Spect 34(1):237–263

Del Vecchio C, Ludovico MD, Prota A (2020) Repair costs of reinforced concrete building components: from actual data analysis to calibrated consequence functions. Earthq Spect 36(1):353–377. https://doi.org/10.1177/8755293019878194

De Martino G, Di Ludovico M, Moroni PA, C, Manfredi G, Dolce M, (2017) Estimation of repair costs for RC and masonry residential buildings based on damage data collected by post-earthquake visual inspection. Bull Earthq Eng 15(4):1681–1706

Di Ludovico M, Prota A, Moroni C, Manfredi G, Dolce M (2017a) Reconstruction process of damaged residential buildings outside historical centres after the L'Aquila earthquake: part I—" light damage" reconstruction. Bull Earthq Eng 15(2):667–692

Di Ludovico M, Prota A, Moroni C, Manfredi G, Dolce M (2017b) Reconstruction process of damaged residential buildings outside historical centres after the L'Aquila earthquake: part II— "heavy damage" reconstruction. Bull Earthq Eng 15(2):693–729

Erdik M (2000) Report on 1999 Kocaeli and Düzce (Turkey) Earthquakes. Boğaziçi Üniversitesi, Kandilli Rasathanesi ve Deprem Araştırma Enstitüsü

FEMA 306 (1998) Evaluation of earthquake damaged concrete and masonry wall buildings: basic procedures manual. ATC, Redwood City, CA, USA

Grünthal G (1998) European macroseismic scale. European Seismological Commission (ESC)

Ilki A, Demir C, Comert M, Kusunoki K (2013) Evaluation of seismic damage assessment methodologies based on field observations, test results and analytical studies. In: 2013 International Van earthquake symposium. (keynote lecture)

Kazama M, Noda T (2012) Damage statistics (Summary of the 2011 off the Pacific Coast of Tohoku Earthquake damage). Soils Found 52(5):780–792

Marquis F, Kim JJ, Elwood KJ, Chang SE (2017) Understanding post-earthquake decisions on multi-storey concrete buildings in Christchurch, New Zealand. Bull Earthq Eng 15(2):731–758

New Zealand Society for Earthquake Engineering (NZSEE) (2009) Building safety evaluation during a state of emergency—Guidelines for territorial authorities

Taghavi S, Miranda E (2003) Response assessment of nonstructural building elements. PEER report 2003/05. Berkeley: University of California

The Japan building disaster prevention association (JBDPA) (2015) Guideline for post-earthquake damage evaluation and rehabilitation (in Japanese)

Part II
Loss Modelling and Insurance Pricing

Chapter 6
Earthquake Risk Assessment from Insurance Perspective

M. Erdik

Abstract The assessment of earthquake and risk to a portfolio, in urban or regional scale, constitutes an important element in the mitigation of economic and social losses due to earthquakes, planning of immediate post-earthquake actions as well as for the development of earthquake insurance schemes. Earthquake loss and risk assessment methodologies consider and combine three main elements: earthquake hazard, fragility/vulnerability of assets and the inventory of assets exposed to hazard. Challenges exist in the characterization of the earthquake hazard as well as in the determination of the fragilities/vulnerabilities of the physical and social elements exposed to the hazard. The simulation of the spatially correlated fields of ground motion using empirical models of correlation between intensity measures is an important tool for hazard characterization. The uncertainties involved in these elements and especially the correlation in these uncertainties, are important to obtain the bounds of the expected risks and losses. This paper looks at the current practices in regional and urban earthquake risk assessment, discusses current issues and provides illustrative applications from Istanbul and Turkey.

6.1 Introduction

In UNISDR terminology, "Risk" is defined as "the combination of the probability of an event and its negative consequences", and "Risk assessment" is defined as "a methodology to determine the nature and extent of risk by analyzing potential hazards and evaluating existing conditions of vulnerability that together could potentially harm exposed people, property, services, livelihoods and the environment in which they depend".

Earthquake risk can be defined as the probable economic, social and environmental consequences of earthquakes that may occur in a specified period of time and is

M. Erdik (✉)
Bogazici University, Kandilli Observatory and Earthquake Research Institute, Istanbul 34684, Turkey
e-mail: erdik@boun.edu.tr

© The Author(s) 2021   111
S. Akkar et al. (eds.), *Advances in Assessment and Modeling of Earthquake Loss*,
Springer Tracts in Civil Engineering,
https://doi.org/10.1007/978-3-030-68813-4_6

determined by using earthquake loss modeling procedures. In this context, the loss is the reduction in the value of an asset due to earthquake damage and risk is the quantification of this loss in terms of its probability (or uncertainty) of occurrence. In simpler terms, the "loss" is the reduction in value of an asset due to damage and the "risk" represents the uncertainty of this "loss".

Earthquakes, which have annually caused an average of USD 34.7 billion in damages (Munich 2016), are one of the most destructive natural perils and can lead to severe economic, social and environmental impacts. Rapid urbanization and the accumulation of assets in seismic areas have led to an increase of earthquake risk in many parts of the world. The 2011 Great East Japan Earthquake was the costliest earthquake with USD 210 billion in economic losses followed by the Hanshin-Awaji earthquake (Kobe earthquake) in 1995 with USD 100 billion in economic losses (Munich 2016). Similarly, loss estimates from a 7.8 magnitude earthquake in Southern California would cause over USD 200 billion in economic losses (USGS 2008).

Public and private enterprises analyze their portfolio of assets to assess and to manage their earthquake risk. In calculating the earthquake risk of each asset, social and economic losses, due to not only physical damage to buildings and facilities but also to the non-structural damage, consequential damage and business interruption are considered. In insurance terminology, these risk assessments and estimations are called as the Catastrophe (or simply, "Cat") Modeling. Insurance companies use these cat models for insurance pricing, portfolio management, to monitor their capital requirements and solvency and to determine their reinsurance needs. Cedents can use the cat models to assess the appropriate structure of their outwards program and to compare technical prices of outwards treaties to market prices.

The components of earthquake risk estimation can be addressed following the modular structure of the HAZUS methodology (Whitman et al. 1997; Kircher et al. 2006; FEMA 2003) illustrated in Fig. 6.1.

For a given inventory of elements (location and physical characteristics) exposed to seismic hazard, the important ingredients of this earthquake risk estimation flowchart are Ground Motion, Direct Physical Damage, Induced Physical Damage, and Direct/Indirect Socio-Economic Losses.

Almost all earthquake risk assessment schemes rely on the quantification of the earthquake shaking as intensity measure parameters using probabilistic or deterministic earthquake hazard models. For a given ground motion (intensity measure) the direct physical damage is determined by the fragility/vulnerability relationships that provide the probability of damage/loss, conditional on the level of intensity measure. Each step of the process incorporates stochastic or random variation associated with all aspects of the modeled phenomenon. Consequently, the earthquake risk estimations should consider the uncertainties in these steps.

In 1990, under the UN-IDNDR (International Decade for Natural Disaster Reduction) program the RADIUS (Risk Assessment Tools for the Diagnosis of Urban Areas against Seismic Disasters) project promoted the earthquake risk assessment and mitigation in the international scale (UNISDR 2000). One of the most used methodologies of earthquake risk assessment originate from HAZUS (www.fema.gov/hazus) where, HAZUS-MH MR4 is a damage- and loss-estimation software

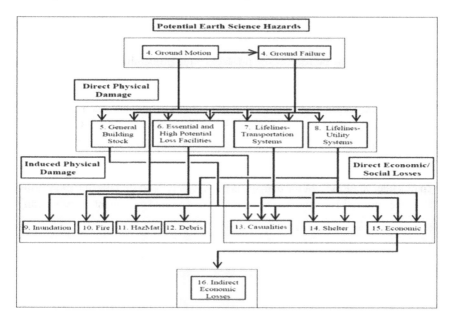

Fig. 6.1 Earthquake risk estimation (after HAZUS-MH, FEMA 2003)

developed by FEMA to estimate potential losses from natural disasters. The World Bank's CAPRA (http://www.ecapra.org/) project has also developed the widely used probabilistic risk assessment tools and software. Besides, several European Projects have also contributed to the development of comprehensive methodologies and tools for earthquake-risk assessment. In this regard, the following projects can be cited: RISK-UE (Mouroux and LeBrun 2006); LESSLOSS (Calvi and Pinho 2004; Spence 2007, http://www.risknat.org/baseprojets/ficheprojet.php?num=55&name= LESSLOSS); SYNER-G (Pitilakis et al. 2014a, b, http://www.vce.at/SYNER-G/ files/dissemination/deliverables.html) and; NERA (www.nera-eu.org).

The Seismology and Earthquake Engineering Research Infrastructure Alliance for Europe (SERA, http://www.sera-eu.org/en/home/) as a Horizon 2020-supported program, works to develop a comprehensive framework for seismic risk modelling at European scale. This risk modeling involves: European capacity curves, fragility, consequence and vulnerability models; European seismic risk results in terms of average annual loss (AAL), probable maximum loss (PML), and risk maps in terms of economic loss and fatalities for specific return periods and; Methods and data to test and evaluate the components of seismic risk models.

GEM initiative (www.globalquakemodel.org), which started in 2006 to develop global, open-source earthquake risk assessment software and tools, has contributed profoundly to the earthquake hazard and risk assessment standards, developed guidelines, the OpenQuake (www.globalquakemodel.org/openquake) software and the global earthquake hazard and risk maps (https://www.globalquakemodel.org/gem).

6.2 Probabilistic Earthquake Risk

Risk can generally be defined as the product of the probability of occurrence of a certain hazard with a prescribed intensity times the consequences of the asset being damaged due to that event. The simple direct way of making probabilistic estimates of damage D exceeding a damage level, d, for a given earthquake: is to express it as a function of the earthquake source, E, and the site, S, parameters (McGuire 2004).

$$P(\text{damage exceeds d|earthquake}) = P(D > d|E, S) \tag{6.1}$$

In practice, since most of the damage is caused by ground shaking, the probability of D (i.e. seismic risk) is estimated as a function of a ground motion Intensity Measure (IM)

$$P(D > d) = \int P(D > d|IM) \times d\lambda(IM > im) \tag{6.2}$$

where:

$P(D > d|IM)$ represents the so-called fragility function and; $\lambda(IM > im)$ is the total frequency, which IM exceeds an intensity measure level "im" and, essentially, represents the seismic hazard at the site.

Yücemen (2013) has developed a discrete for the calculation of risk, in terms of Expected Annual Damage Ratio or Average Annual Loss Ratio ($AALR_k$) for a given (kth) element of the inventory exposed to earthquake hazard.

$$AALR_k = \sum_{IM} MDR_k(IM) \times \lambda(IM)MDR_k(IM) \times \lambda(IM) \tag{6.3}$$

where $MDR_k(IM)$ is the Mean Damage Ratio associated with the inventory element k for the given IM, essentially representing the discrete fragility function, and $\lambda(IM)$ is the total frequency for given IM, essentially representing the seismic hazard.

$$MDR_k = \sum_{DS} P_k(DS|IM \times CDR_k(DS)) \tag{6.4}$$

where $P_k(DS|IM)$ represents the probability for the given inventory element k at a given Damage State (DS) and constitutes the element of the Damage Probability Matrix (DPM) for the inventory element k and $CDR_k(DS)$ represents the Central Damage Ratio for the given inventory element k at the given DS. The DPM, for a given inventory element k, provides the damage probability distribution for different DS (represented by CDR) and the IM.

The development of Performance-Based Earthquake Engineering (PBEE) has created a rigorous and comprehensive framework for Probabilistic Seismic Risk

Analysis (PSRA) (Cornell and Krawinkler 2000; Krawinkler 2002). This PBEE-PSRA framework is based upon a chain of four conditional random variables: the ground motion intensity measure (IM); the engineering demand parameter (EDP), the component-specific damage measure (DM or damage state DS) and, the decision variable (DV). The IM term is a quantitative measure of ground motion shaking intensity such as peak ground acceleration or spectral displacement. The EDP term is a quantitative measure of peak demand on the asset (e.g. inter-story drift ratio, peak floor acceleration for a building). The DS term represents a discrete component damage state. The Decision Variables, DV, is the outcome of the earthquake risk (such as the annual earthquake loss or the exceedance of damage limit states). These parameters are and need to be carefully defined. For example, an efficient IM should be able to predict EDPs with low uncertainty.

Estimation of the DVs involve the assessment of earthquake ground motion, analysis of the structural response and comparison of the response parameters with the performance objectives (Cornell and Krawinkler 2000). In PBEE-PSRA, the annual rate of the DV is provided by total probability integral (so-called, triple integral) provided in Eq. 6.5.

$$\lambda(DV) = \int\limits_{DM} \int\limits_{EDP} \int\limits_{IMG} (DV|DM) \times dG(DM|EDP) \times dG(EDP|IM) \times d\lambda(IM) \qquad (6.5)$$

where: $\lambda(DV)$ is the annual rate of exceeding the decision variable, DV;

G(DV|DM) is the probability of exceeding the decision variable given the damage measure, DM;

G(DM|EDP) is the probability of exceeding the damage measure, DM, given the engineering demand parameter, EDP;

G(EDP|IM) is the probability of exceeding the engineering demand parameter, EDP, given the intensity measure, IM, and;

$\lambda(IM)$ is the annual rate of exceeding the ground motion intensity measure and;

dG(DV|DM), dG(DM|EDP)λ and dλ(IM) are the differentials of the respective terms.

The steps used in the PBEE-PSRA, as indicated in the total probability integral given by Eq. 6.5 are illustrated in Fig. 6.2 (Moehle 2003).

Following processes can be distinguished in Eq. 6.5 and in Fig. 6.2:

- Hazard Analysis represents the annual rate of exceedance of certain intensity measures (IMs), where λ(IM) quantifies the annual rate of exceeding a given value of seismic intensity measure (IM) (i.e. the outcome of the PSHA).
- In the structural analysis, one creates a structural model of the building in order to estimate the response, measured in terms of a vector of engineering demand parameters (EDP), conditioned on seismic excitation represented by a set of IMs [G(EDP|IM)].
- Damage Analysis yields the conditional probability function, G(DM|EDP), that relates Damage Measures (DMs) and EDP. The DM distributions are generally characterized in terms of fragility curves.

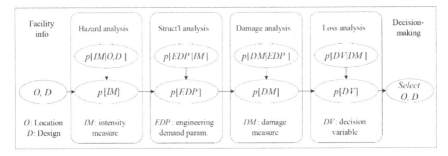

Fig. 6.2 Steps in the PBEE-PSRA procedure (Moehle 2003)

- Loss Analysis uses Decision Variable (DV) as the random variable and produces the conditional probability function, G(DV|DM), for given DMs, to describe the earthquake risk (e.g. the annual losses, the exceedance of damage limit states).

In Eq. 6.5 all four variables (IM, EDP, DM, and DV) are continuous random variables. However, Eq. 6.5 is generally modified as the summation of discrete terms, since in the current practice; the damage measures are not continuous but rather a set of discrete damage states. The integration of scenario losses provided by the triple integral (Eq. 6.5) over the entire range of occurrence probability will result in the quantification of seismic risk in terms of the Expected Annual Loss (EAL) (Dhakal and Mander 2006).

6.2.1 Fragility Functions

In general, seismic fragility is defined as the probability that the damage of a structure exceeds a specific damage state "d" for a given level of seismic hazard (McGuire 2004)

$$\text{Fragility} = P \{\text{Damage} > d \mid \text{Seismic Hazard}\} \tag{6.6}$$

Melchers (1999) provides the following expressions to define the general fragility functions.

$$FR(x) = P(\text{Loss}|IM = x) \tag{6.7}$$

and

$$\lambda(\text{Loss}) = \int_x FR(x) \times d\lambda(IM) \tag{6.8}$$

where FR(x) denotes the fragility function for a specific loss for a given IM = x and λ(Loss) is the annual rate of exceedance of the specific Loss.

In PBEE-PSRA, the fragility functions are assigned for discrete damage states, to provide the probability of exceeding a damage state for a given EDP level, as shown in Eq. 6.7.

$$\lambda(EDP) = \int_{IM} G(EDP|IM) \times d\lambda(IM) \qquad (6.9)$$

where, λ(EDP) is the annual rate of exceeding a specified demand level EDP$\geq$edp; G(EDP|IM) is the probability of exceeding the engineering demand parameter, EDP, given the intensity measure, IM and; λ(IM) is the annual rate of exceeding the ground motion intensity measure, IM. For the assessment λ(EDP), the result of probabilistic seismic demand analysis can be used.

The conditional distribution G(EDP|IM) can also be called "Demand Fragility Function". Similarly, the "Damage Fragility Function" and "Loss Fragility Function" corresponding, respectively to DM and DV can be derived as follows (Lu et al. 2012):

$$G(DM|IM) = \int_{EDP} G(DM|EDP) \times dG(EDP|IM) \qquad (6.10)$$

$$G(DV|IM) = \int_{DM} \int_{EDP} G(DV|DM) \times dG(DM|EDP) \times dG(EDP|IM)$$

$$(6.11)$$

These equations can be further reduced to yield respectively the annual rate of exceeding a specified damage measure level (DM $\geq$ dm) and decision level (DV $\geq$ dv).

$$\lambda(DM) = \int_{IM} G(DV|IM) \times d\lambda(IM) \qquad (6.12)$$

$$\lambda(DV) = \int_{IM} G(DM|DM) \times d\lambda(IM) \qquad (6.13)$$

6.3 Ground Motion Intensity Measures (IM)

Estimates of damage to structures are made on the basis of a given level of ground motion intensity. The strength of an earthquake ground motion is often quantified by an IM (Baker and Cornell 2005). Macroseismic intensity and peak ground motion parameters (e.g. peak ground acceleration, velocity, and displacement, PGA, PGV and PGD, respectively) as well as the spectral acceleration/displacement at the fundamental vibration period of the structure, have been traditionally used in earthquake vulnerability assessment studies (Calvi et al. 2006). The use of a particular intensity measure for fragility or vulnerability assessment depends on the damageability characteristics of the element under the direct and indirect actions induced by an earthquake.

6.3.1 Ground Motion Prediction Models

Ground-motion predictive models (GMPMs) provide a probability distribution for ground motion intensity measures and are modeled in the following form (Baker 2013)

$$\ln(IM) = \left(\overline{\ln(IM)}\right)(M, R, \Theta) + \sigma(M, R, \Theta) \cdot \varepsilon \qquad (6.14)$$

where ln(IM) is the logarithm of ground motion intensity measure that is modeled as a normally distributed random variable. The terms $\overline{\ln(IM)}(M, R, \Theta)$ and $\sigma(M, R, \Theta)$ are the predicted mean and standard deviation of the ln(IM), respectively. They are functions of magnitude, M, source-site distance, R and other estimator parameters such as rupture mechanism, soil conditions and etc. that are collectively referred in vector Θ. The parameter ε is a standard normal random variable and represents the variability in ln(IM). Positive ε produces larger than-average values of ln(IM), whereas negative ε values yield smaller-than-average values of ln(IM).

Given a ground-motion intensity measure of interest the exceedance probability of any *im* level is computed from the predicted mean $\left(\left(\overline{\ln(IM)}(M, R, \Theta)\right)\right)$ and standard deviation ($\sigma(M, R, \Theta)$) as given below.

$$P(IM >; im|m, r, \theta) = 1 - \Phi\left(\frac{\ln(im) - \overline{\ln IM}}{\sigma \ln IM}\right) \qquad (6.15)$$

here, $\Phi()$ is standard normal cumulative distribution. Equation (6.13) can be alternatively written in the form of probability distribution function ($f_{IM}(u)$) as in Eq. (6.14) that is generally more convenient for PSHA.

$$P(IM > im|m, r, \theta) = \int\limits_{im}^{\infty} \frac{1}{\sigma_{lnIM}\sqrt{2\pi}} \exp\left(-\frac{1}{2}\left(\frac{\ln(u) - \overline{\ln(IM)}}{\sigma_{lnIM}}\right)^2\right) du \quad (6.16)$$

Following Jayaram and Baker (2009): the logarithm of a ground motion Intensity Measure, IM_{ij}, at a site i for an earthquake j, is modeled from Eq. (6.12).

$$\ln\left(IM_{ij}\right) = \overline{\ln\left(IM_{ij}\right)}\left(M_j, R_{ij}, \Theta_{ij}\right) + \sigma_{ij}\varepsilon_{ij} + \tau_j\eta_j \tag{6.17}$$

The standard deviation in Eq. 6.14 is now decomposed into two components: σ_{ij} and ε_{ij} describe the within-event (inter-event) variability and, τ_j and η_j describe the between-event (intra-event) variability. σ_{ij} and τ_j are intra-event and inter-event standard deviations, respectively, ε_{ij} is the normalized intra-event residual at site i for earthquake j and η_j is the normalized inter-event residual for earthquake j. The total residual is the sum of inter- and intra-event residuals and the total standard deviation σ_{Ti} is given by Eq. 6.18.

$$\sigma_{T_{ij}} = \sqrt{\sigma_{ij}^2 + \tau_j^2} \tag{6.18}$$

6.3.2   Spatial Correlation of Ground Motion

It has been shown that, for a given earthquake, spatial correlation of IMs exists and it is essentially attributable to the following two sources:

(1)   The event-wide correlation of IMs through the between-event (intra-event) variability (i.e. a systematic lower or higher ground motion of an event, for instance, due to a higher or lower stress drop at the source) and:

(2)   The tendency of local IMs being lower or higher than the GMPMs predicted median, through the within-event (intra-event) variability (i.e. near-fault directivity effects and wave propagation paths). (e.g. Wang and Takada 2005; Goda and Hong 2008a, b; Jayaram and Baker 2009; Esposito and Iervolino 2011). The intra-event residuals at different sites are correlated, as a function of their separation distance. This correlation would be larger as the distance between the sites become smaller.

As such, when modeling ground motion fields for a scenario earthquake, a sample of the inter-event residuals for all the sites/cells, for the event, should be taken and combined with the intra-event residual at each site/cell, obtained through a spatial correlation model (e.g. Crowley et al. 2008).

For the estimation of the spatial correlation of ground motion IMs, semivariogram definitions are generally used (Jayaram and Baker 2009). The ηj (normalized inter-event residual for earthquake j) is a random variable that follows a standard normal distribution. The σ_{ij} (normalized intra-event residual at site i for earthquake j) is a 2-dimensional random field that exhibits spatial correlation following a model semivariogram, defined as a measure of the average dissimilarity in data values as a function of the separation distance between sites. Jayaram and Baker (2009) proposed the semivariogram, $\rho(h)$, for predicting spatial correlation of ground motions IMs with a given separation distance h.

Fig. 6.3 A typical semivariogram

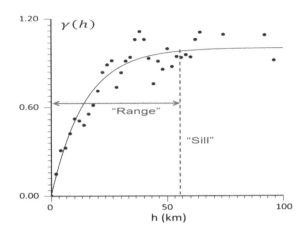

$$\gamma(h) = a\big[1 - \exp(-3h/b)\big] \qquad (6.19)$$

where: a is the sill and b is the range of the semivariogram function. Both a and b factors depend on the IM calculated and the geological conditions (Fig. 6.3).

It is generally sufficient to estimate the semivariogram in order to estimate the correlation coefficient, since it can be theoretically shown that the following relationship holds between the semivariogram, γ (h), and the correlation coefficient, $\rho(h)$, (Goovaerts 1997):

$$\gamma(h) = a(1 - r(h)) \qquad (6.20)$$

Empirical models of spatial correlation of ground motion intensity measures exist only for a few seismic regions in the world, such as Japan, Taiwan, California and Marmara Region in Turkey, since a dense observation of strong earthquake ground motion is necessary for this purpose.

Wagener et al. (2016) has studied the intra-event correlation of the earthquake ground motion using the data from Istanbul Rapid Response and Early Warning System (Erdik et al. 2003). A comparison of the different correlation models is provided in Fig. 6.4 (Wagener et al. 2016). As it can be assessed, a large variability in the correlation lengths (1.65–43.5 km) exist. Correlation models based on Japanese and Taiwanese data generally exhibit a more gradual decay with distance and longer correlation lengths compared to Marmara and the California models. The period dependence of the PSA correlation coefficient (more gradual decay with distance for longer periods) was observed by Goda and Hong (2008a, b) in California as well as by Goda and Atkinson (2010) in Japan.

For conducting spatially correlated simulations of ground motion, two of the utilized formulations are the covariance and spectral formulations (Wu and Baker 2014). In Covariance Formulation, the intra-event residual term, ε_{ij}, can be constructed through sampling from a multivariate normal distribution with a {0}

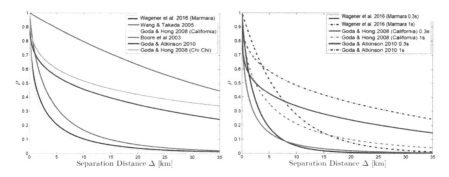

Fig. 6.4 Comparison of different correlation models reported in the literature (Wagener et al. 2016)

mean vector and covariance matrix, Σ_{xy}, defined by Eq. 6.21 (Wu and Baker 2014).

$$\sum_{xy} 1 - \gamma\left(h_{xy}\right) = e\left(-\frac{3h_{xy}}{b}\right) \tag{6.21}$$

where, Σ_{xy} is the covariance matrix between sites x and y, stored as the (xth, yth) element of the n × n (n is number of sites/cells) covariance matrix Σ. Covariance formulation has a significant computational expense and can only be used for a modest number of sites.

6.3.3 Correlation Between IMs at the Same Site

If different IMs are needed for different fragility or vulnerability functions (especially for heterogeneous assets) these IMs need to be cross-correlated. The cross-correlation between the residuals of spectral accelerations (the difference between SA(T1) and SA(T2) based on a GMPM) feature a correlation that depends on the inter-period difference (Baker and Cornell 2006). Several models have been proposed for correlation of pseudo-spectral accelerations recorded at the same site (Inoue and Cornell 1990; Baker and Cornell 2006; Baker and Jayaram 2008). Plots of the correlation coefficients based on the predictive equation provided in Baker and Jayaram (2008) are given in Fig. 6.5.

Weatherill et al. (2013) provides a comprehensive description of the generation random fields of ground shaking considering the inter-period correlation of the ground motion residuals.

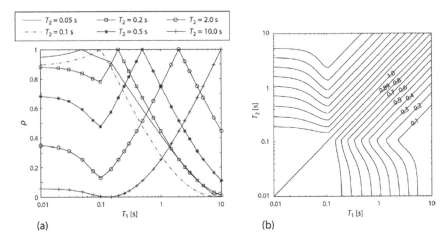

Fig. 6.5 Plots of correlation coefficients (Baker and Jayaram, 2008). **a** Plots of correlation coefficients versus T1, for several T2 values; **b** contours of correlation coefficients versus T1 and T2

6.4 Probabilistic Seismic Hazard Assessment (PSHA)

PSHA is an essential part of Probabilistic Seismic Risk Analysis (PSRA) in the Performance-Based Earthquake Engineering (PBEE) framework (McGuire 2004; Goulet et al. 2007; Ruiz-Garcia and Miranda 2007). Hazard Curves provide the probabilities of exceedance in a given time period for given values of an IM. Probabilistic Seismic Hazard Assessment (PSHA) is used to obtain the annual rate of exceeding the ground motion intensity measure, $\lambda(IM)$. PSHA is traditionally represented by following the total probability theorem (Cornell 1968; McGuire 2004):

$$\lambda(IM > x) = \lambda(M > m_{min}) \int_{m_{min}}^{m_{max}} \int_{0}^{r_{max}} P(IM > x|m, r) f_M(m) f_R(r) dr \, dm$$

(6.22)

where: $\lambda(M > m_{min})$ is the annual rate of earthquakes with magnitude greater than or equal to the minimum magnitude m_{min}, r is the source distance, m is the earthquake magnitude and; $f_M(m)$ and $f_R(r)$ are the probability density functions (PDFs) for the magnitude and distance. It should be noted that: this equation is indicated for a single earthquake source zone and the integration is over all considered magnitudes and distances. The integration process can be extended to encompass other earthquake sources as well.

6.4.1 Monte Carlo Simulation

Monte-Carlo method can be utilized to estimate the probabilistic seismic hazard, instead of the computation through the total probability integral given by Eq. 6.22. The same also holds for probabilistic seismic risk applications through the total probability integral given by Eq. 6.5. As such, for seismic hazard and risk assessment applications, it is rational to carry out a numerical evaluation of the probabilistic earthquake risk using a Monte Carlo simulation-based approach.

Monte Carlo method is based on a multitudinous resampling of an earthquake catalog to construct synthetic earthquake catalogs and then to find earthquake ground motions from which the hazard values are found (Ebel and Kafka 1999). For PSHA assessment first element of the Monte Carlo simulation technique is to generate synthetic earthquake catalogs (Stochastic Event Sets) for each source zone by drawing random samples from the assumed PSHA model components (Musson 2000; Scherbaum et al. 2004). Subsequently, the ground motion intensity-measures (IMs) can be evaluated for each earthquake contained in the catalog and, for all earthquakes in the catalog, a history of ground motion IM estimates is obtained at each site. These estimates are reorganized to develop a list of the annual maximum IMs in ascending order, to yield the seismic hazard curve through a plot of the sorted annual maximum IMs as a function of the probability of exceedance.

6.4.2 Ground Motion Distribution Maps

Ground-motion IM Field Maps describes the geographic distribution of a given IM obtained considering an earthquake rupture and a GMPM. The spatial correlation of the intra-event residuals can be considered in the generation of the field.

The earthquake shaking can be determined theoretically for assumed (scenario) earthquake source parameters through median ground motion prediction models or, for post-earthquake cases, using a hybrid methodology that corrects the analytical data with empirical observations. These type of maps are generally called as "ShakeMaps" (Wald et al. 2006, https://earthquake.usgs.gov/data/shakemap/). In insurance industry, post-earthquake ShakeMaps are used with industry exposure data to calculate insured loss estimates (Parametric Earthquake Insurance).

For the analysis of seismic risk (especially for distributed assets), it is necessary to produce a spatially correlated field of ground motion. The ground motion IM across a region should be defined in a manner that is consistent with either a given earthquake scenario or a given return period. Pitilakis et al. (2014b) refers to these maps as a "Shakefield".

6.4.3 Risk-Based Earthquake Hazard: Risk-Targeted Hazard Maps for Earthquake Resistant Design

The earthquake resistant design of structures requires the definition of design basis ground motion for a given return period, with the assumption that the probability of collapse for buildings is uniform regardless of the location. However, for a rigorous and explicitly uniform probability of collapse, the hazard maps should essentially be risk-based. In ASCE-SEI codes, risk target is taken as 1% probability of collapse in 50 years).

The distribution of the collapse capacity in terms of a specific IM can be defined by a cumulative lognormal function with log mean, β, and log standard deviation, β. Luco et al. (2007), using a $\beta = 0.8$, found a probability of collapse of 10% at 2475-year ground motion level in the USA. Douglas et al. (2013), using a $\beta = 0.5$, found a probability of collapse of 10–5 at the 475-year return period design ground motion level, for new buildings in France.

6.5 Assets Exposed to Earthquake Hazard, Building Inventories

Assets Exposed to Hazard is represented by the Exposure Model that contains the information regarding the assets (such as building inventories) within the area of interest for the assessment of earthquake risk.

To perform a seismic risk assessment, building inventories are determined based on specific classification systems (taxonomies).

Building taxonomies define structure categories by various combinations of use, time of construction, construction material, lateral force-resisting system, height, applicable building code, and quality (FEMA 2003; EMS-98-Grunthal 1998 and RISK-UE 2004).

Publicly available data, at country and regional spatial scale, includes: UN-Housing database, UN-HABITAT, UN Statistical Database on Global Housing, Population and Housing Censuses of individual Countries (https://en.wikipedia.org/wiki/Population_and_housing_censuses_by_country), the World Housing Encyclopedia (WHE) database developed by EERI and IAEE (http://www.world-housing.net).

The "Global Exposure Database for the Global Earthquake Model" project, under the Global Earthquake Model (GEM—www.globalquakemodel.org) framework, is concerned with the compilation of an inventory of assets at risk (Gamba et al. 2014 and Dell'Acqua et al. 2012). The USGS—Prompt Assessment of Global Earthquakes for Response (PAGER) (Wald et al. 2010) undertaking has also developed a comprehensive global inventory of assets exposed to earthquake hazard (Jaiswal et al. 2010).

6.6 Fragility, Consequence and Vulnerability Relationships

Fragility Relationships (Models) describe the probability of exceeding a set of damage states, given an intensity measure level. HAZUS (FEMA 1999) uses four damage states as the: slight, moderate, extensive and complete damage. Combining the fragility information with consequence (damage to loss) functions, which describe the probability distribution of loss given a performance (damage) level, allows for the derivation of vulnerability functions. Vulnerability functions can be used to directly estimate economic losses, where the loss ratio could be the ratio of cost of repair to the cost of replacement for a given building typology.

The seismic fragility and vulnerability functions are customarily modeled by lognormal cumulative distribution functions (CDF) (Ellingwood 2001; Wen et al. 2004). The fragility function is the conditional probability of reaching or exceeding a specific value of the Damage State (DS >= ds) or a Damage Index (DI) reaching or exceeding a specific value of Limit State (LS) (DI => LS), for a given Intensity Measure (IM). On the other hand, the vulnerability function is represented by the conditional probability of reaching or exceeding a specific value of the loss (Loss >= loss) for a given Intensity Measure (IM), such as Spectral Displacement (Sd). By assuming a lognormal distribution of EDP at a given IM, the fragility function, defined as the damage index (DI) compared with the limit states (LS), corresponding to various damage states (SD) for given intensity measure (IM), can be written as:

$$P[DI \geqslant LS|IM] = 1 - \Phi\left(\frac{\ln(LS) - \ln(EDP)}{\sigma_{EDP|IM}}\right) \quad (6.23)$$

where, $\sigma_{EDP|IM}$ is the standard deviation of the logarithmic EDP distribution given by $EDP = a (IM)^b$ or $\ln(EDP) = \ln a + b \ln(IM)$, (a and be are regression coefficients resulting from the response data) and; $\Phi(\bullet)$ is the standard normal distribution function.

In more colloquial terms: the fragility function, which determines the probability that a building or a group of buildings will be in, or exceed, the ith damage state, d_i, for a given level of ground motion intensity (IM), is commonly expressed conditionally in the form of a cumulative lognormal distribution.

$$P[d \geqslant d_i|IM] = 1 - \Phi\left(\frac{\ln(IM) - \ln(IM_{di})}{\beta_{di}}\right) \quad (6.24)$$

where P is the exceedance probability for the ith damage state, IM is the ground motion intensity, Φ is the standard normal cumulative distribution function, β_{di} is the standard deviation of the natural logarithm (dispersion) of ground motion intensity for the ith damage state and IM_{di} is the median value of ground motion intensity at which the subject reaches the ith damage state. Typical fragility and vulnerability functions are illustrated in Fig. 6.6, after Luco and Karaca (2006).

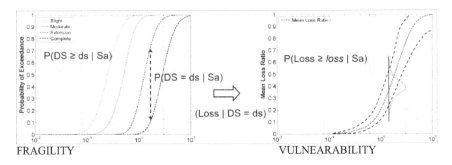

Fig. 6.6 Typical fragility and vulnerability functions (Luco and Karaca 2006)

Numerous approaches exist towards "direct" estimation of fragility and vulnerability functions at various levels of resolution. Approaches that are generally used for the "direct" estimation fragility and vulnerability functions are empirical, analytical and hybrid.

Empirical fragility and vulnerability relationships based on the use of macro-seismic intensity (MMI, EMS'98) and the observed damage/loss data is still a valuable choice for fragility and vulnerability modeling. An empirical vulnerability function for Turkish building stock is provided in Fig. 6.7 to provide an example. Using the EMS'98 (Grunthal 1998) intensity definitions, Giovinazzi and Lagomarsino (2004) developed a method on the basis of beta damage distribution and fuzzy set theory to produce fragility matrices. This method has been incorporated into the ELER earthquake loss assessment tool.

Analytical (or predicted) fragility refers to the assessment of the expected performance of buildings based on calculation and building characteristics. The capacity

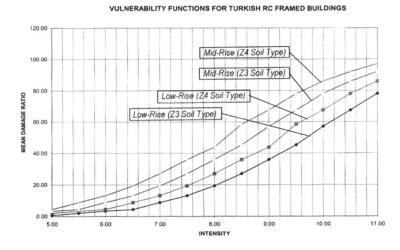

Fig. 6.7 Intensity based empirical vulnerability relationship for Turkish RC framed buildings

Table 6.1 Mean damage ratios for EMS'98 damage grades

Damage grade	OYO (2009)	HAZUS (1999)	Bramerini et al. (1995)	ATC-13 (1987)	Tyagunov et al. (2006)
D1—No damage	0.05	0.02	0.01	0.05	0.05
D2—Slight damage	0.2	0.1	0.1	0.2	0.1
D3—Moderate damage	0.5	0.5	0.35	0.55	0.4
D4—Extensive damage	0.8	1	0.75	0.9	0.8
D5—Complete damage	1	1	1	1	1

spectrum method, originally derived by Freeman (1998), is first implemented within the HAZUS (FEMA 2003) procedure as well as in many other earthquake loss estimation analyses (e.g. SELENA—Molina and Lindholm, 2010 and ELER (Erdik et al. 2008; Hancılar et al. 2010). DBELA (Displacement-Based Earthquake Loss Assessment) method (Crowley et al. 2004; Bal et al. 2008) relies on the principles of direct displacement-based design method of Priestley (2003).

Vulnerability functions can also be derived '"indirectly" through the combination of a fragility function and a damage-to-loss or consequence models. Consequence models provide the values of "Mean Damage Ratio (MDR)", defined as the cost of repairing the structure (to bring it to its pre-earthquake state) divided by its replacement cost, for each damage state. A typical table is provided in Table 6.1 for EMS'98 damage grades (Yepes-Estrada et al. 2014).

Several compilations of literature-based fragility and vulnerability functions exist. Such as: GEM database of vulnerability and fragility functions for buildings (Yepes-Estrada et al. 2016; Yepes-Estrada et al. 2014) and SYNER-G database for infrastructure fragilities (Pitilakis et al. 2014b; Crowley et al. 2014).

Correlation of Vulnerability/Fragility Uncertainties

In general, fragility and vulnerability function correlations are incorporated only for limit cases of independent or perfectly correlated component damage states and, it is generally not possible to do more than an estimate the losses, with and without vulnerability uncertainty correlation, to constrain the results. Evidence of correlation of vulnerability and fragility function uncertainties can be obtained from post-earthquake damage surveys.

6.7 Metrics Used in Risk Assessment and CAT Modeling

For the measurement of risk for a single asset or portfolio of assets, several metrics, in physical and financial loss terms, are used. Following is a brief explanation of these metrics.

The Loss Exceedance (or Exceedance Probability, EP) curves, the Average Annual Loss (AAL) and Probable Maximum Loss (PML) constitute the primary metrics of the probabilistic risk/loss assessment. In engineering terms, the losses associated with the building stock are generally quantified in terms of Los Ratio (LR), defined as the repair cost divided by the replacement cost. LR is also called as the damage factor, damage ratio and fractional loss.

Loss Exceedance Curves (EP Curves) describe losses versus probability of exceedance in a given time span (generally, annual). EP Curves are used for cat modelling, as it is beneficial to identify attachment or exhaustion probabilities, calculate expected losses within a given range, or to provide benchmarks for comparisons between risks or over time.

Occurrence Exceedance Probability (OEP) is the probability that the associated loss level will be exceeded by any event in any given year. It provides information on losses assuming a single event occurrence in a given year. Aggregate Exceedance Probability (AEP) is the probability that the associated loss level will be exceeded by the aggregated losses in any given year. It provides information on losses assuming one or more occurrences in a year.

The AEP and OEP can be used for managing exposure both to single large event and to multiple events across a time period. They can be similar when the probability of two or more events is very small; they are identical when there is zero probability of two or more events. However, AEP can be very different from the OEP when the probability of two or more events is significant.

Value at Risk (VaR) is equivalent to the Return Period, and measures a single point of a range of potential outcomes corresponding to a given confidence. The VaR is the fractile value on an EP curve corresponding to a selected probability level.

Tail Value at Risk (TVaR) measures the mean loss of all potential outcomes with losses greater than a fixed point. When used to compare two risks, along with mean loss and Value at Risk, it helps communicate how quickly potential losses tail off.

VaR and TVaR are both mathematical measures used in cat modelling to represent a risk profile, or range of potential outcomes, in a single value.

Conditional Value at Risk (CVaR) is the area under the EP curve below a selected cumulative probability level, p, normalized by the probability of exceedance $(1 - p)$. CVaR, accounts for the rare events in terms of their severity and frequency by taking the conditional expectation of the EP curve.

Average Annual Loss (AAL) (or Annual Estimated Loss—AEL or Pure Premium) is the expected value of a loss exceedance distribution and can be computed as the product of the loss for a given event with the probability of at least one occurrence of event, summed over all events. AAL is the average loss of all modeled events, weighted by their probability of annual occurrence (EP curve) and corresponds to the area underneath the EP curve. If the loss ratio (LR) is used for the quantification of loss, then the term Average Annual Loss Ratio (AALR) is used in lieu of AAL. For earthquake insurance purposes, the AAL or AALR is of particular importance in determining the annual pure premiums.

Pure Premium represents the average of all potential outcomes considered in the analysis, and could be considered to be the break-even point if such a policy is to be written for very large number of times.

The Probable Maximum Loss (PML) is one of the most popular metrics in financial risk management, and there are several definitions. PML can be associated with the OEP or the AEP. Conventionally, PML was defined as a fractal of the loss corresponding to the return period of 475 years. In Japan, the PML is defined as the (conditional) 0.9-fractile value for a scenario that corresponds to a selected probability level (typically, return period of 475 years).

ASTM E2026-16A use specific nomenclatures for seismic risk assessment of buildings. are in use:

Scenario Upper Loss, based on deterministic analysis) (SUL) is defined as the earthquake loss to the building with a 90% confidence of non-exceedance (or a 10% probability of exceedance), resulting from a specified event on specific faults affecting the building. If the specified earthquake hazard is the 475-year return period event, then this term can be called the SUL475, and this term is the same measurement as the traditional PML defined above.

Scenario Expected Loss, based on deterministic analysis, (SEL) is defined as the average expected loss to the building, resulting from a specified event on specific faults affecting the building. If the specified earthquake is the 475-year return period event, then this term can be called the SEL475.

The Probable Loss, based on probabilistic analysis, (PL) is defined as the earthquake loss to the building(s) that has a specified probability of being exceeded in a given time period from earthquake shaking. The PL is commonly taken as the loss that has a 10% probability of exceedance in 50 years, which is called the PL475, because it corresponds to a return period of 475 years.

6.8 Earthquake Risk Assessment Models and Example Applications

The estimation of the earthquake risk due to deterministic earthquake scenarios is of use for communicating seismic risk to the public and to decision makers. However, a probabilistic assessment of earthquake risk (generally called, Probabilistic Seismic Risk Analysis-PSRA) is needed for risk prioritization, risk mitigation actions and for decision-making in the insurance and reinsurance sectors.

Seismic risk for a single element at risk can be calculated through the convolution of a hazard curve with a vulnerability relationship quantifying the probability of a given consequence occurring under different levels of ground shaking. For geographically distributed elements, the use of hazard curves calculated with conventional PSHA, may overestimate the total loss since the conventional PSHA does not distinguish the inter- and intra-earthquake variability of ground motion (Crowley and Bommer 2006).

Since the PSRA encompasses multitude sources of uncertainties stemming from hazard, inventory and vulnerability (or fragility and consequence) functions, Monte Carlo simulations are routinely employed to facilitate the orderly propagation of these uncertainties within the process. Using Monte Carlo simulations, a value of the inter-earthquake variability can be sampled for each earthquake and then values of the intra-earthquake variability are sampled at each location for this earthquake. Such event-based simulation involves suites of probabilistically characterized deterministic risk scenarios (e.g. Crowley and Bommer 2006; Silva et al. 2013).

Similar to PSHA, the results of a PSRA can also be deaggregated to identify the components of the overall system (i.e. earthquake scenarios) that are contributing significantly to the seismic risk (e.g. Goda and Hong 2008a, b; Jayaram and Baker 2009).

One of the first rational assessment of earthquake risk is carried out by Whitman et al. (1973) using MMI versus Damage Ratio matrices. Similar studies in USA led to the development of HAZUS (FEMA 2003) Some of the open access and state-of-the-art software packages for earthquake risk assessment can be listed as follows:

- CAPRA GIS-Earthquake module, http://www.ecapra.org/software
- EQRM, http://www.ga.gov.au/scienti.ic-topics/hazards/earthquake/capabilties/modelling/eqrm
- ERGO (MAEviz/mHARP), http://ergo.ncsa.illinois.edu/?page id = 48
- HAZUS-MH earthquake module, http://www.fema.gov/hazus
- OpenQuake, https://www.globalquakemodel.org/openquake/
- ELER,http://www.koeri.boun.edu.tr/Haberler/NERIES%20ELER%20V3.16 176.depmuh
- RiskScape-Earthquake, https://riskscape.niwa.co.nz/
- SELENA, http://www.norsar.no/seismology/engineering/SELENA-RISe/
- EQVIS, http://www.vce.at/SYNER-G/files/downloads.html

The OpenQuake Engine (https://www.globalquakemodel.org/) is GEM 's state-of-the-art software for seismic hazard and risk assessment at varying scales of resolution, from global to local. It is open-source, fully transparent and can be used with GEM or user-developed models to carry out scenario-based and probabilistic hazard and risk calculations and produce a great variety of hazard and loss outputs. Spatial correlation of the ground motion residuals and correlation of the uncertainty in the vulnerability can be modeled. Main calculations performed in connection with the earthquake loss assessment can be listed as: Scenario risk; Scenario damage; Classical PSHA-based risk; Probabilistic event-based risk and; Retrofitting benefit-cost ratio. Comprehensive global earthquake risk maps were provided by GEM (https://www.globalquakemodel.org/gem).

Today, the seismic risk/loss assessment can be essentially grouped under the following three approaches (Silva et al. 2013):

- Deterministic Risk/Loss Calculation (analysis due to a single earthquake scenario);
- Probabilistic Risk/Loss Calculation (an analysis that considers a probabilistic description of the earthquake events and associated ground motions) and;
- Classical PSHA-Based Risk/Loss Calculation (analysis based on conventional probabilistic earthquake hazard assessment).

6.8.1 Deterministic Earthquake Risk/Loss Calculation

In this approach, the earthquake loss is computed due to a single, deterministic earthquake scenario. The flowchart of the process is shown in Fig. 6.8. The hazard input consists of a scenario earthquake source parameters and a single (or a set of) GMPMs. By repeating the same deterministic loss assessment process, and sampling the inter- and intra-variability (spatial variability) from the GMPM each time, many ground motion fields can be computed to account for the aleatory variability in the ground motion. The losses to all assets (with different typologies and vulnerabilities) in the region of interest can be aggregated per ground motion field, for statistical studies of aggregated losses.

Following are some earthquake risk assessment examples, where, deterministic earthquake loss calculation procedure is used.

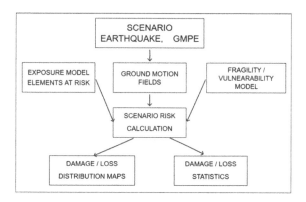

Fig. 6.8 Simplified flowchart of the Deterministic Earthquake Loss Calculation process

6.8.1.1 Deterministic Loss Assessment for Buildings in a Region in Istanbul

In recent studies (i.e. Bohnhoff et al. 2013; Ergintav et al. 2014) the Princess Islands Segment of the Main Marmara Fault (Fig. 6.9) has been identified as the "most imminent danger" to Istanbul. This fault segment has been considered (Type: Strike-Slip; Magnitude: Mw7.3, Recurrence Model: Fully Characteristic; Slip Rate: 20 mm/year, Dip: 90°, Rake: 0°) with the Kale et al. (2015) GMPM and spatial correlation models of Wagener et al. (2016) and Goda and Hong (2008a) to compute 1000 simulations of earthquake ground motion distribution and then the loss in the most densely populated region of Istanbul.

Figures 6.10 and 6.11 illustrate the median and 84-percentile PGA and PGV shake fields after 1000 simulations of ground motion based on Kale et al. (2015) GMPM and Wagener et al. (2016) correlation models.

For loss assessment the intensity-based fragility relationships of Lagomarsino and Giovinazzi (2006) are considered. Instrumental intensities were computed from PGA and PGA-conditioned PGV distributions using the Wald et al. (1999) relationships. Figure 6.12 provides the median and 84-percentile loss ratios (for mid-rise,

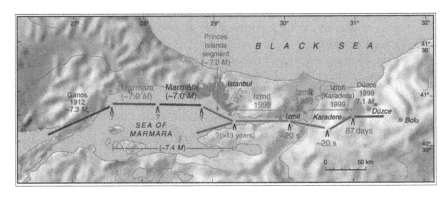

Fig. 6.9 Princess Islands Segment of the Main Marmara Fault (Bohnhoff et al. 2013)

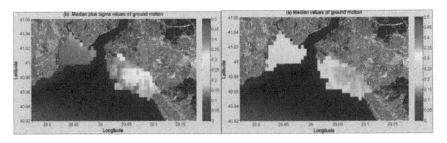

Fig. 6.10 Median (right panel) and 84 percentile (left panel) Shake Fields, PGA (g), 1000 Simulations

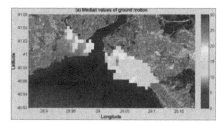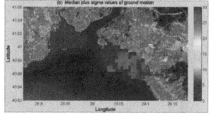

Fig. 6.11 Median (left panel) and 84 percentile (right panel) Shake Fields, PGA-conditioned PGV (cm/s), 1000 Simulations

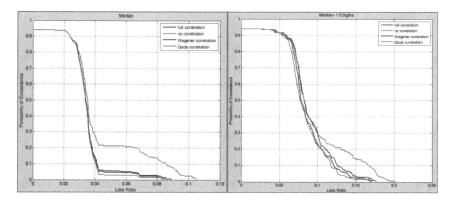

Fig. 6.12 Median (left) and 84-percentile (right) loss ratios (includes all building types considered in the database) for full and no spatial correlation of intra-event residuals and Wagener et al. (2016) and Goda and Hong (2008a, b) relationships-based spatial correlations

post-2000 R/C buildings) for cases of full and no spatial correlation of intra-event residuals and for the cases of Wagener et al. (2016) and Goda and Hong (2008a, b) relationships-based spatial correlations. The same spatial correlation cases were also used in Fig. 6.13, where the total economic loss curve (Exceedance Probability-EP Curve) due to structural damage is illustrated in Fig. 6.13. As it can be assessed: for high probabilities (low loss) spatial correlation results are above full correlation results (similar to no correlation results) and; for low probabilities (high loss) spatial correlation results are located between the full correlation and no correlation results. Full correlation and no correlation cases overlap each other because of the intra-event residuals are zero mean normal distributions and these marginal cases would yield the same results since under any mathematical operation (linear or nonlinear) realized for a zero-mean normal variable will yield a zero mean variable. Figure 6.12 provide distributions of the loss ratio in the geographic region considered corresponding to median and median + sigma intensity distributions. Figure 6.13 approaches the same problem probabilistically and provides the likelihood of experiencing different total economic loss values in the region considered by taking into account the annual occurrence frequency of each of the 1000 events simulated.

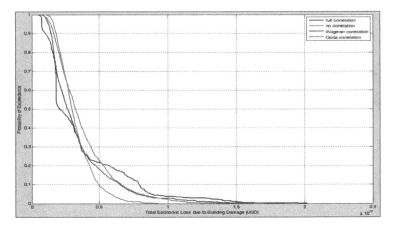

Fig. 6.13 Total economic loss curve (Exceedance Probability-EP Curve) due to structural damage to all building types considered in the database, for full and no spatial correlation of intra-event residuals and Wagener et al. (2016) and Goda and Hong (2008a, b) relationships-based spatial correlations

6.8.1.2 Deterministic Earthquake Loss Assessment in the Zeytinburnu District of Istanbul

Wagener et al. (2016) has computed the losses in the Zeytinburnu district of Istanbul that would result from an Mw7.2 scenario earthquake on the Marmara Fault. The building portfolio consisted of 11,250 reinforced concrete and masonry buildings. PGA, SA(0.3 s) and SA(1 s) distributions were calculated considering various spatial correlation models as well as their cross-correlations. The GMPM relationship of Akkar and Bommer (2010) was considered. Figure 6.14 illustrate the effects of the different correlation models on the spatial distribution of PGA. The loss histograms are shown in Fig. 6.15 with the distribution parameters, mean μ, median m, standard

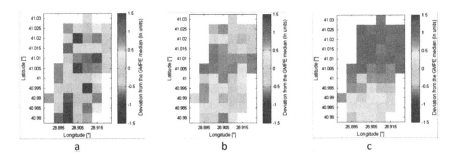

Fig. 6.14 Realizations of simulated PGA-distribution in Zeytinburnu with various correlation properties. **a** No spatial correlation, **b** Wagener et al. (2016) correlation model **c** a simple one-parameter exponential decay with 20 km correlation length

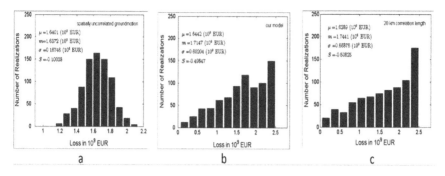

Fig. 6.15 Histograms of aggregated economic loss in Zeytinburnu with various correlation properties. **a** No spatial correlation, **b** Wagener et al. (2016) correlation model **c** a simple one-parameter exponential decay with 20 km correlation length

deviation σ, and skewness S. As it can be seen, while the mean loss remains essentially unaltered, the coefficient of variation increases with increasing correlation from 0.102 to 0.411. From Fig. 6.15, it can be assessed that the uncorrelated ground motion (a) results in a narrow, bell-shaped loss distribution with a mean of 1.62 billion € and a coefficient of variation (CV) of 0.108. When correlation models are implemented (b, c), the mean loss remains unaltered but the shape of the loss distribution changes significantly. The coefficient of variation increases and the loss distribution is skewed towards higher losses. The included spatial correlation increases the likelihood of simultaneous large ground motions in many geocells, as illustrated in the realizations of simulated PGA-distributions in Fig. 6.14. Consequently, the likelihood of damage to many buildings increases. If the correlation lengths increase with the spectral period, mid- and high-rise buildings will be more affected by correlation properties than low-rise buildings, since higher variability in the loss distributions, were observed when increasing the correlation length.

Deterministic Earthquake Risk Assessments for İstanbul

Silva et al. (2012) provides an application of Openquake software-based deterministic loss calculation to develop a mean loss (USD) distribution map for İstanbul as shown in Fig. 6.16. The application considers a deterministic Mw7.5 magnitude earthquake on the Main Marmara Fault. Only reinforced concrete buildings are considered, which constitutes about 80% of the building inventory (Erdik 2010). Boore and Atkinson (2008) GMPM was used to produce about 500 ground motion fields, where the intra-event variability was sampled by the model proposed by Jayaram and Baker (2009).

A comprehensive earthquake risk assessment study was conducted by Boğaziçi University, OYO International and GRM Ltd. for İstanbul Metropolitan Municipality (İBB) in 2009 (http://istanbul-olasi-deprem-kayiplari-tahminlerinin-guncel lenmesil_sonuc_rapor_2010.09.pdf). This study was updated in 2018 by Boğaziçi University for the İBB (http://depremzemin.ibb.istanbul/wp-content/uploads/2020/02/dezim_kandilli_deprem-hasar-tahmin_raporu.pdf). The 2009 study was based on

Fig. 6.16 Distribution of
mean economic losses in
Istanbul (Silva et al. 2012)

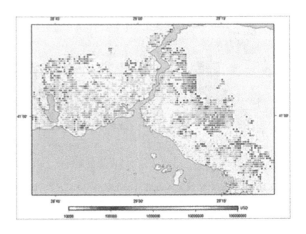

a single scenario earthquake (Mw7.5), rupturing the Main Marmara Fault to simulate
the so-called pending "İstanbul Earthquake" with an annual probability of occurrence
of about 2–3%. Intensity- and spectral acceleration-based fragilities were considered.
Loss ratios for the buildings, as well as other losses, were determined for median and
84-percentile probabilities. In addition to this Mw7.5 scenario earthquake, the 2018
study also considered single stochastic ground motion simulations for several rupture
alternatives and the official PSHA map for different return periods, for the earthquake
ground motion. The aggregate building damage results for different damage states,
obtained from different rupture scenarios, do not differ much from the results of the
Mw7.5 scenario earthquake. The risk in both studies was computed using a classical
simple deterministic approach, with no consideration of spatial variation of ground
motion intensity.

The building damage rates that would result from the occurrence of the Mw 7.5
Istanbul earthquake scenario indicate that the median damage ratios for buildings with
no, light, medium, heavy and very heavy/collapse damage status are respectively
found to be about 60, 26,11, 2 and 1%. Noting that as of 2020 there are about
1.1 million buildings and 3.9 million housing units in the İstanbul Province, one
can estimate that about 0.44 million buildings (about 1.6 million housing units)
will receive some degree of damage after exposure to the "İstanbul Earthquake". In
monetary terms, this structural damage will correspond to about USD 6.5 billion.

On the basis of the 2009 Boğaziçi Uninersity risk assessment study, Hancılar
et al. (2010) provides for the distribution of the rates of medium building damage
in İstanbul (Fig. 6.17) in terms of the number of damaged buildings per geographic
cell. Distribution of building structural loss ratio is provided in Fig. 6.18, using the
site-specific median ground motion for a deterministic scenario earthquake of Mw7.5
on the Main Marmara Fault.

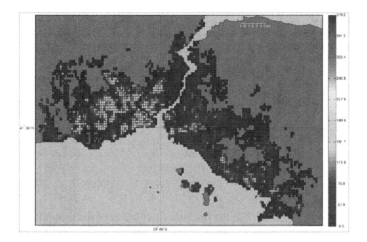

Fig. 6.17 Rates of building damage in İstanbul for medium damage grade (Median ground motion from deterministic scenario earthquake of Mw7.5 on the Main Marmara Fault) (Hancılar et al. 2010)

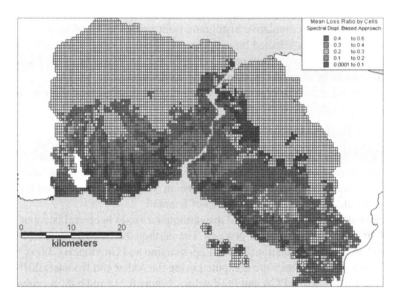

Fig. 6.18 Distribution of building structural loss ratio in İstanbul (Median ground motion from deterministic scenario earthquake of Mw7.5 on the Main Marmara Fault)

6.8.2 Probabilistic Earthquake Risk Calculation

In the Probabilistic Loss Calculation process the probability of losses and loss statistics are computed using Monte-Carlo simulations, based on stochastic event sets and associated ground motion fields Goda and Yoshikawa (2012). The flowchart of the

Fig. 6.19 Simplified
flowchart of the Probabilistic
Loss Calculation process

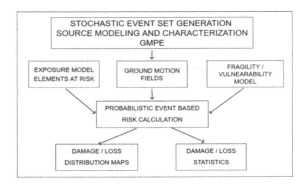

process is shown in Fig. 6.19. For the realistic calculation of the ground motion field for each event, the sampling of the inter-event variability and the spatial correlation of the intra-event residuals of the ground-motion model should be considered. The set of ground-motion fields are combined with the exposure and vulnerability model to obtain losses. In this combination, the correlation of the uncertainty in the vulnerability model may also be incorporated. For the computation of the loss exceedance curve: the cumulative histogram, built using the list of losses per asset (of a given typology) in selected bins of loss over the time span, can be considered. An aggregated loss curve, representative of the whole set of assets within the region (or portfolio) can be obtained by aggregating all the losses.

Following are some earthquake risk assessment examples, where, probabilistic earthquake loss calculation procedure is used.

Probabilistic Earthquake Risk Assessments for İstanbul

Using GEM OpenQuake Probabilistic Loss Calculation process, Crowley et al. (2011) present (respectively in Fig. 6.20a and b) a loss map and a total loss exceedance curve for a probability of exceedance of 10% in 50 years for reinforced concrete buildings located in the metropolitan area of Istanbul.

Akkar et al. (2016) have computed the earthquake losses in central Istanbul using the probabilistic loss calculation process. The earthquake risk was assessed using intensity-based fragility relationships of Lagomarsino and Giovinazzi (2006), where the instrumental intensities were computed using the Akkar and Boomer (2010) and Akkar et al. (2014)-based PGA and PGV values. Figure 6.21a and b show probability exceedance of very heavy damage state (DS4 in EMS'98) in 50 years for post-2002 low-rise RC buildings.

Turkish Catastrophe Insurance Pool (TCIP) Loss Modeling

In 2018 TCIP has engaged Turkish Earthquake Foundation (TDV) to carry out an estimation of earthquake losses to its insured portfolio losses for reinsurance purposes. A comprehensive investigation encompassing a review of existing building census, building types, soil conditions, near-fault effects, spatial distribution of ground motion, fragility relationships and consequence functions was conducted.

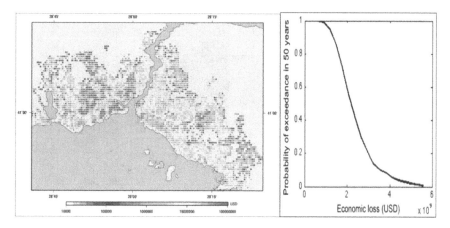

Fig. 6.20   a and b Loss map and loss exceedance curve for a probability of exceedance of 10% in 50 years for RC buildings located in metropolitan Istanbul (Crowley et al. 2011)

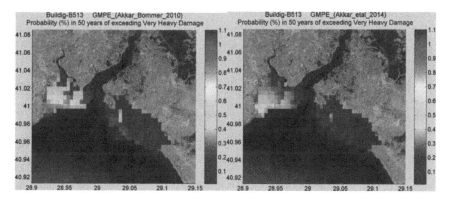

Fig. 6.21   a and b Probability exceedance of very heavy damage state (DS4 in EMS'95) in 50 years for post-2002 low-rise RC buildings built after 2002. Akkar and Bommer (2010) **a** and Akkar et al. (2014) **b** GMPMs are used

Two types of loss calculation procedures are considered: (A) Monte Carlo process based SHA considering all of the seismic sources in the region 5000 × 100-year catalogs generated and (B) Monte Carlo process based SHA considering only the selected segments of the Main Marmara Fault (i.e. North Anatolian Fault) that are believed to be primed for rupture for an annual probability of about 2–3% (Fig. 6.22). On these segments 2000 deterministic earthquake scenario models of magnitude between Mw6.8–7.4 and hypcentral depths between 0 and 18 km (simulating the so-called pending Istanbul Earthquake) are generated.

To provide some examples of the results, the loss of the TCIP portfolio insured value (Procedure—A) for the Istanbul Province (EP Curve) is provided in Fig. 6.23. The vertical axes of this EP curve show annual probability of exceedance and also

Fig. 6.22 Istanbul province with the 0.05° × 0.05° geo-cells and the segments of the Main Marmara Fault considered in the loss assessment procedure

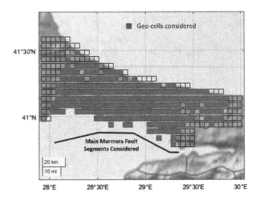

Fig. 6.23 Loss of TCIP portfolio insured value for the Istanbul Province (Procedure-A on thle left and Procedure-B on the right)

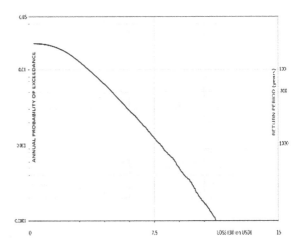

the average return period. The loss of the TCIP insured value for 100- and 200-year average return periods are approximately 3 and 5 Billion USD, respectively.

Figure 6.24 shows the loss of the TCIP portfolio insured value (Procedure—B) for the Istanbul Province (EP Curve). The vertical axis of this EP curve shows the probability of exceedance, since the deterministic earthquake scenarios considered in the analysis refer to different models of the same (the pending Istanbul Earthquake) event. The loss of the TCIP portfolio insured value is about USD 3 Billion USD (median) with a standard deviation of USD 2 Billion. The median ratio of buildings that will receive no, light, medium, heavy and very heavy/collapse damage status are respectively found to be about 40%, 41%, 10%, 8% and 1%. Noting that as of 2020 the number of TCIP policies in İstanbul Province is about 2.5 million, one can expect, on the average, about 1.5 million claims only from the İstanbul Province, in the event of the "İstanbul Earthquake".

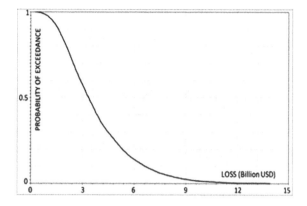

Fig. 6.24 Loss of TCIP portfolio insured value for the Istanbul Province (Procedure-A on thie left and Procedure-B on the right)

6.8.3 Classical PSHA-Based Earthquake Risk Calculation

In this approach, classical PSHA assessment (Cornell 1968; McGuire 2004) can be used to calculate loss exceedance curves for single assets, calculated site by site, on the basis of hazard curves. The flowchart of the process is shown in Fig. 6.25. Discrete vulnerability functions are converted into a loss exceedance matrix (e.g. a matrix which describes the probability of exceedance of each monetary loss value or loss ratio for a discrete set of intensity measure levels). The values of each column of this matrix are multiplied by the probability of occurrence of the associated intensity measure level, extracted from the hazard curves. To compute the loss exceedance curve: the probabilities of exceedance of the loss (or the loss ratio) curve are obtained by summing all the values per loss (or loss ratio).

Demircioğlu et al. has computed the grid-based building damage distributions, loss ratios (LR) and average annual loss ratios (AALR) corresponding to 72, 475, and 2475-year average return periods. Figure 6.26 provides sub-province based LR

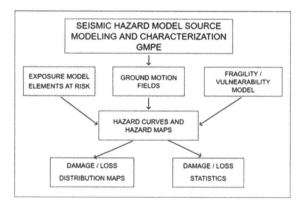

Fig. 6.25 Simplified flowchart of the classical PSHA-based loss calculation

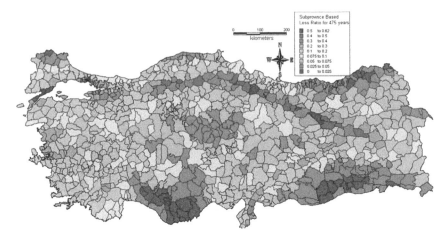

Fig. 6.26 Sub-province based loss ratios for 475-year average return period

values for Turkey for the 475-year average return period. Figure 6.27 provides geo-cell based LR values for the Marmara Region for the 475-year average return period at geo-cell ($0.05° \times 0.05°$) resolution.

A classical PSHA-based earthquake loss assessment for California have been carried out using HAZUS (FEMA 2003) to estimate county-based Annual Economic Loss and Annual Average Loss Ratios (Fig. 6.28, Chen et al. 2016). Similar studies have also been conducted for Turkey using ELER for the assessment of sub-district based AAL values (Fig. 6.29).

Studies of the Global Earthquake Model (GEM) initiative have culminated in 2018 in the development of global earthquake risk maps (https://www.globalquakemodel.org/gem). Figures 6.30 and 6.31 provides respectively the AALR maps and the EP

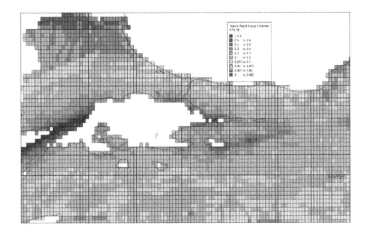

Fig. 6.27 Geo-cell based loss ratios for 475-year average return period

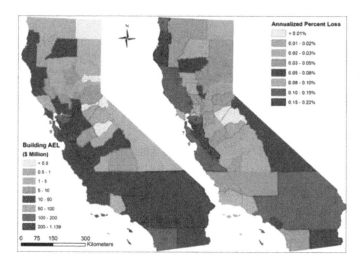

Fig. 6.28 County-based annual economic loss and annual average loss ratios for California (Chen et al. 2016)

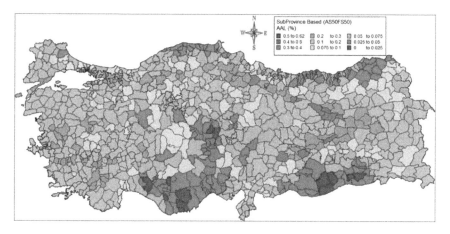

Fig. 6.29 Sub district-based annual average loss ratios for Turkey

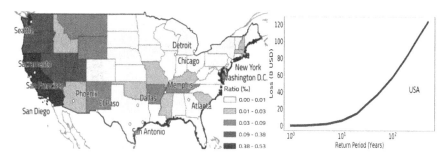

Fig. 6.30 AALR distribution and EP curve for the USA (https://www.globalquakemodel.org/gem)

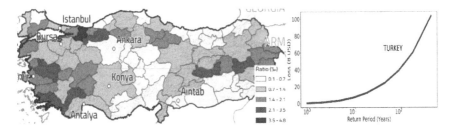

Fig. 6.31 AALR distribution and EP curve for Turkey (https://www.globalquakemodel.org/gem)

curve for the USA and Turkey. In these figures the average annual loss ratio represents the long-term mean average annual loss normalized by the total asset replacement cost within the subdivision due to direct damage caused by earthquake ground shaking in the residential, commercial and industrial building stock, considering structural and non-structural components and building contents.

Turkish Catastrophe Insurance Pool (TCIP) AALR Models

In 2018 TCIP has engaged Turkish Earthquake Foundation (TDV) to carry out a risk based update of its insurance pricing in consideration of the newly prepared national earthquake hazard map and the prevailing building typology. A comprehensive investigation encompassing a review of existing building census, building typology, fragility relationships and consequence functions was conducted. To provide some examples of the results. the AALR distribution map for all buildings on sub-district basis is provided in Fig. 6.32 and the geo-cell (0.05° × 0.05°) based distribution of AALR for post 1979, low rise reinforced concrete frame buildings in the Marmara Region of Turkey is shown in Fig. 6.33.

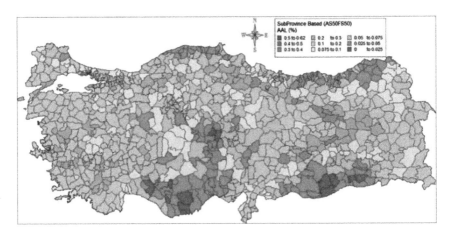

Fig. 6.32 Sub-district based distribution of AALR (%) for the total building stock (After TCIP-TDV project)

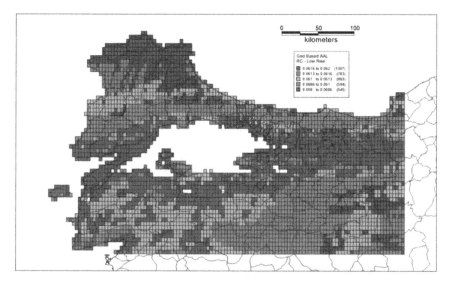

Fig. 6.33 Geo cell (0.05° × 0.05°) based distribution of AALR for post 1979, low rise reinforced concrete frame buildings in the Marmara Region of Turkey (After TCIP-TDV project)

6.8.4 Effect of the Spatial Correlation of Ground Motion on Earthquake Loss Assessments

The effect of the consideration of spatial correlation of IMs, can be assessed from the examples provided for both deterministic and probabilistic earthquake loss applications: Sect. 6.8.1.1 Deterministic Loss Assessment for Buildings in a Region in Istanbul and Sect. 6.8.1.2 Deterministic Earthquake Loss Assessment in the Zeytinburnu District of Istanbul. The findings in these sections essentially follow those obtained by Park et al. (2007) who has performed stochastic simulation of ground motion fields to compute seismic losses within two portfolios of structures. Annual Mean Rate of Exceedance (essentially, EP) curves, for building portfolios with large and small footprints are assessed for six different models for the correlation coefficient, varying from no correlation at all distances to fully correlated ground motion fields, to study their effect on the EP curves. Park et al. (2007) has observed that: for either portfolio type, no correlation related losses associated with low probabilities of exceedance are significantly underestimated compared to the cases with correlation. The relative underestimation of losses associated with low probabilities of exceedance are evident for portfolios with small footprint than that with the large footprint and the effect of spatial correlation on the entire portfolio was found to be larger if the correlation length is comparable or larger than the footprint of the portfolio.

6.9 Uncertainties in Risk Assessments

The main sources of uncertainties in earthquake risk assessment are:

- Hazard uncertainty (seismic source characterization and ground motion modeling)
- Vulnerability uncertainty
- Uncertainty in the assumptions and specifications of the risk model
- Portfolio uncertainty (location and other attributes of the building classes).

In general, there exist two types of uncertainties that need to be considered in earthquake risk/loss assessments: aleatory and epistemic. Aleatory uncertainty accounts for the randomness of the data used in the analysis and the epistemic uncertainty accounts for lack of knowledge in the model.

Aleatory variability, that generally affects the loss distributions and exceedance curves is directly included in the probabilistic analysis calculations through the inclusion of the standard deviation of a GMPM considered in the analysis. Epistemic uncertainties, which can increase the spread of the loss distributions, are generally considered by means of a logic tree formulation with appropriate branches and weights associated with different hypotheses. Similarly, Monte-Carlo techniques can also be used to examine the effect of the epistemic uncertainties in loss estimates.

Demand Surge and Loss Amplification represent the so-called Post Event Inflation elements in earthquake risk assessment. They arise due to: Shortages of labor and materials, which cause prices to rise; Supply/demand imbalances delay repairs, which results in structural deterioration and; Political issues (due to the size of the disaster and under pressure from politicians, insurers are encouraged to settle claims generously).

Figure 6.34 (after Wong et al. 2000) illustrates the effect of uncertainties on loss estimation. Uncertainties arise in part from incomplete inventories of the built environment, inadequate scientific knowledge of the process, earthquake ground motion (IMs) and their effects upon buildings and facilities (fragility/vulnerability relationships). The reliability of the fragility/vulnerability relationships is essentially related to the conformity of the ground motion IMs with the earthquake performance (damage) of the building inventory. These uncertainties can result in a range of uncertainty in loss estimates, at best, a factor of two.

The general finding of the studies on the uncertainties in earthquake loss estimation is that the uncertainties are large and at least as equal to uncertainties in hazard analyses (Stafford et al. 2007; Strasser et al. 2008). It should also be noted that the estimates of human casualties are derived by uncertain relationships from already uncertain building loss estimates, so the uncertainties in these estimates are rather compounded (Coburn and Spence 2002).

Financial loss caused by the earthquakes is, essentially, the translation of physical damage into total monetary loss using local estimates of repair and reconstruction costs. Several regression-based simplified equations are developed to calculate earthquake losses. The failure of such simple procedures that stem from the extensive uncertainties in the physical process are exemplified in Fig. 6.35 (after Daniell 2014)

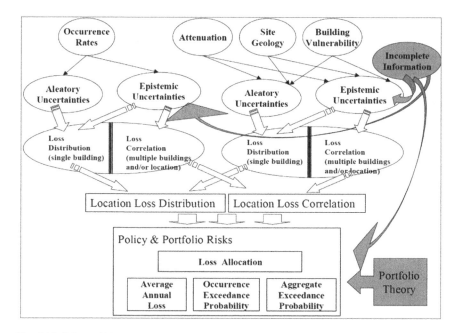

Fig. 6.34   Effect of Uncertainties on Loss Estimation (Wong et al. 2000)

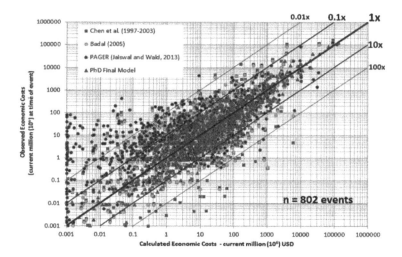

Fig. 6.35   Observed versus calculated costs for 4 different studies (After Daniell 2014)

where a comparison between the observed and calculated financial losses caused by earthquakes are shown. As it can be seen, the inherent uncertainties in the loss calculations can cause differences up to two orders of magnitude between the observed and calculated financial losses.

To show the effect of these uncertainties on the AALR distributions, Fig. 6.36 provides a comparison of sub-district based AALR distribution in Turkey, prepared by TCIP and by different vendor Cat-Models, Although the AALR values of the vendor models are not shown, the colors from red, pink, yellow, light green to dark green indicate decreasing values of AALR. The difference in the distribution of these colors between the four models is significant and evidences the effect of data and modeling uncertainties. Another evidence of these uncertainties is illustrated in Fig. 6.37, where a comparison of Exceedance Probability (EP) curves for Istanbul

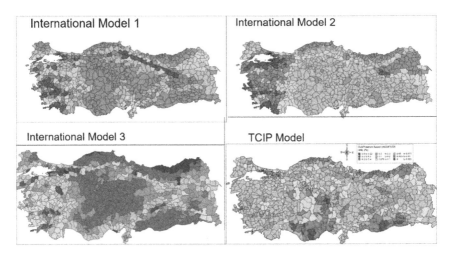

Fig. 6.36 Comparison of AALR distribution prepared by TCIP and by different vendor Cat-Models (The AALR values of the vendor models are not shown. The colors from red, pink, yellow, light green to dark green indicate decreasing values of AALR)

Fig. 6.37 Comparison of Exceedance Probability Curves for Istanbul by different vendor Cat-Models

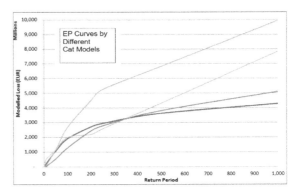

by different vendor Cat-Models are provided. As it can be observed differences up to 100% exist. These EP curves cannot be directly compared with Fig. 6.23 due to the much smaller (about one-half) TCIP portfolio used in their computation.

6.10   Conclusions

- Earthquake risk and loss assessment is needed to prioritize risk mitigation actions, emergency planning, and management of related financial commitments. Insurance sector have to conduct the earthquake risk analysis of their portfolio to assess their solvency in the next major disaster, to price insurance and to buy re-insurance cover.
- Due to the research and development on rational probabilistic risk/loss assessment methodologies and studies conducted in connection with several important projects, today we have substantial capability to analyses the risk and losses ensuing from low-probability, high consequence major earthquake events.
- In this regard, the selection of an appropriate set of GMPMs, that are compatible with the regional seismo-tectonic characteristics, and the selection of vulnerability (or fragility and consequence) relationships that are compatible with the IMs and appropriate with the inventory of assets in the portfolio are of great importance. The mean damage ratio (MDR) is highly sensitive to the consequence models (i.e. loss ratios assigned to each damage state).
- The probability distribution function for the loss to a portfolio depends on the spatial correlation of the ground motion and the vulnerability of the buildings. The consideration of the spatial correlation does not change the mean loss but increases the dispersion in the loss distribution, which can have a profound influence in loss and insurance related decisions. When spatial correlation is considered, the losses at longer return periods increase. On the opposite side, the losses at shorter return periods may be overestimated if the spatial correlation is not included in the analysis.
- The reduction of the uncertainties in earthquake risk/loss assessment is an important issue to increase the reliability and to reduce the variability between the assessments resulting from different of earthquake risk/loss models. In this connection, earthquake risk/loss assessment models should explicitly account for the epistemic uncertainties in the components of analysis, especially in the inventory of assets and vulnerability relationships.
- The practice of risk assessment is now established. However, a number of research issues, such as: uncertainty correlation in vulnerability, logic-tree modeling of epistemic uncertainties and treatment of uncertainties in exposure modeling, remain for treatment in future applications.

Acknowledgements The valuable contribution of my colleagues (listed in alphabetical order) Prof. Dr. Sinan Akkar, Assoc. Prof. Dr. Zehra Çağnan, Dr. Yin Cheng, Dr. Mine Betül Demircioğlu, Dr. Karin Şeşetyan and Dr. Thomas Wagener, in the preparation of this paper are gratefully acknowledged.

References

Akkar S, Y Cheng, M Erdik (2016) Implementation of Monte-Carlo simulations for probabilistic loss assessment of geographically distributed portfolio using multi-scale random fields: a case study for Istanbul. In: Proceedings of SSA 2016 annual meeting, Reno, Nevada

Akkar S, Bommer JJ (2010) Empirical equations for the prediction of PGA, PGV and in Europe, the Mediterranean and the Middle East. Seism Research Letters 81:195–206

Akkar S, Sandikkaya MA, Bommer JJ (2014) Empirical ground-motion models for point- and extended-source crustal earthquake scenarios in europe and the middle east. Bull Earthq Eng 12:359–387

ASTM E2026–16A Standard guide for seismic risk assessment of buildings

ATC-13 (1985) Earthquake damage evaluation data for California, Report ATC-13, Applied Technology Council, Redwood City, California, U.S.A

Baker JW (2013) Probabilistic seismic hazard analysis. White Paper Version 2.0.1, 79 pp

Baker JW, Cornell CA (2005) A vector-valued ground motion intensity measures consisting of and epsilon. Earthq Eng Struct Dyn 34:1193–1217

Baker JW, Cornell CA (2006) Correlation of response spectral values for multi-component ground motions. Bull Seism Soc Am 96:215–227

Baker JW, Jayaram N (2008) Correlation of values from NGA ground motion models. Earthq Spect 24(1):299–317

Bal IE, Crowley H, Pinho R (2008) Displacement-based earthquake loss assessment for an earthquake scenario in istanbul. J Earthq Eng 11(2):12–22

Bohnhoff M, Bulut F, Dresen G, Malin PE, Eken T, Aktar M (2013) An earthquake gap south of Istanbul. Nat Commun 4:1999

Boore DM, Atkinson GM (2008) Ground-motion prediction equations for the average horizontal component of PGA, PGV, and 5%-damped PSA at spectral periods between 0.01 s and 10.0 s. Earthq Spect 24:99–138

Bramerini F et al (1995) Rischio sismico del territorio Italiano: proposta per una metodologia e risultati preliminary, Technical report, SSN/RT/95/01, Rome, Italy

Calvi GM, Pinho R (2004) LESSLOSS—a European integrated project on risk mitigation for earthquakes and landslides. IUSS Press, Pavia, Italy

Calvi GM, Pinho R, Magenes G, Bommer JJ, Restrepo-Vélez LF, Crowley H (2006) The development of seismic vulnerabilty assessment methodologies over the past 30 years. ISET J Earthq Tech 43(4):75–104

Chen R, Jaiswal KS, Bausch D, Seligson H, Wills CJ (2016) Annualized earthquake loss estimates for California and their sensitivity to site amplification. Seism Research Letters 87 (8)

Coburn A, Spence R (2002) Earthquake protection, 2nd edn. John Wiley and Sons Ltd, Chichester, England

Cornell CA (1968) Engineering Bull Seismo Soc Am 58:1583–1606

Cornell CA, Krawinkler H (2000) Progress and challenges in seismic performance assessment, PEER Center News Spring, 3(2)

Crowley H, Bommer JJ (2006) Modelling seismic hazard in earthquake loss models with spatially distributed exposure. Bull Earthq Eng 4:249–273

Crowley H, Monelli D, Pagani M, Silva V, Weatherill G (2011) OpenQuake Book. GEM Foundation, Pavia

Crowley H, Colombi M, Silva V (2014) Chapter 4: Epistemic uncertainty in fragility functions for European RC buildings. In: Pitilakis K, Crowley H, Kaynia A (eds) SYNER-G: Typology definition and fragility functions for physical elements at seismic risk: buildings, lifelines, networks and critical facilities,

Crowley H, Pinho R, Bommer JJ (2004) A probabilistic displacement-based vulnerability assessment procedure for earthquake. Bull Earthq Eng 2(2):173–219

Crowley H, Bommer JJ, Stafford PJ (2008) Recent developments in the treatment of ground-motion variability in earthquake loss models. J Earthq Eng 12(1):71–80

Daniell JE (2014) The development of socio-economic fragility functions for use in worldwide rapid earthquake loss estimation procedures, Doctoral Thesis (in publishing), Karlsruhe, Germany

Dell'Acqua F, Gamba P, Jaiswal K (2012) Spatial aspects of building and population exposure data and their implications for global earthquake exposure modeling. Nat Hazards 68(3):1291–1309

Dhakal RP, Mander JB (2006) Financial risk assessment methodology for natural hazards. Bull New Zealand Soc Earthq Eng 39(2):91–105

Douglas J, Ulrich T, Negulescu C (2013) Risk-targeted seismic design maps for mainland France. Nat Haz 65

Ebel JA, Kafka AL (1999) A Monte Carlo approach to seismic hazard analysis. Bull Seism Soc Am 89:854–866

Ellingwood BR (2001) Earthquake risk assessment of building structures. Reliab Eng & Syst Safety 74(3):251–262

Erdik M, Fahjan Y, Özel O, Alçık H, Mert A, Gül M (2003) Istanbul earthquake rapid response and early warning system. Bull Earthq Eng 1:157–163

Erdik M (2010) Mitigation of earthquake risk in Istanbul. In: Seismic risk management in urban areas-proceedings of a U.S.-Iran-Turkey Seismic Workshop December 14–16, 2010 Istanbul, Turkey, PEER Report, Berkeley, USA

Erdik M, Cagnan Z, Zulfikar C, Sesetyan K, Demircioglu MB, Durukal E, Kariptas C (2008) Development of rapid earthquake loss assessment methodologies for Euro-MED Region. In: Proc. 14. World Conference on Earthquake Eng, Paper ID: S04-004

Erdik M, Sesetyan K, Demircioğlu M, Hancılar U, Zülfikar C, Çaktı E, Kamer Y, Ye-nidoğan C, Tüzün C, Çağnan Z, Harmandar E (2010) Rapid earthquake hazard and loss assessment for EuroMediterranean Region. Acta Geophysica 58(5):855–892

Ergintav S et al (2014) Istanbul's Earthquake hot spots: geodetic constraints on strain accumulation along faults in the Marmara seismic gap. Geophys Res Lett 41(16):5783–5788

Esposito S, Iervolino I (2011) PGA and PGV spatial correlation models based on European multievent datasets. Bull Seism Soc Am 101(5):2532–2541

FEMA (1999) HAZUS earthquake loss estimation methodology. Technical manual, Prepared by the National Institute of Building Sciences for the Federal Emergency Management Agency, Washington DC

FEMA-154 (2002) Rapid visual screening of buildings for potential seismic hazards: a handbook, federal emergency management agency, Washington. DC

FEMA (2003) HAZUS-MH technical manual. D.C., Federal Emergency Management Agency, Washington

Freeman SA (1998) Development and use of capacity spectrum method. Proceedings of the 6th U.S. National Conference on Earthquake Engineering, Oakland, California, EERI

Gamba et al (2014) Global exposure database: scientific features, GEM Technical Report 2014-10, Global Earthquake Model, Pavia, Italy

Goovaerts P (1997) Geostatistics for natural resources evaluation. Oxford University Press, UK

Giovinazzi S, Lagomarsino S (2004) A macroseismic model for the vulnerability assessment of buildings. In: 13th World Conference on Earthquake Engineering. Vancouver, Canada

Goda K, Hong HP (2008b) Scenario earthquake for spatially distributed structures. In: The 14th world conference on earthquake engineering October 12–17, 2008, Beijing, China

Goda K, Hong HP (2008a) Spatial correlation of peak ground motions and response spectra. Bull Seism Soc Am 98(1):354–365

Goda K, Atkinson GM (2010) Intraevent spatial correlation of ground-motion parameters using SK-net data. Bull Seism Soc Am 100:3055–3067

Goda K, Yoshikawa H (2012) Earthquake insurance portfolio analysis of wood-frame houses in south-western British Columbia. Canada, Bull Earthq Eng 10:615–643

Goulet CA, Haselton CB, Mitrani-Reiser J, Beck JL, Deierlein G, Porter KA et al (2007) Evaluation of the seismic performance of code-conforming reinforced-concrete frame building-From seismic hazard to collapse safety and economic losses. Earthq Eng Struct Dyn 36(13)

Grünthal G (ed.) (1998) European Macroseismic Scale 1998 (EMS-98). Cahiers du Centre Européen de Géodynamique et de Séismologie 15, Centre Européen de Géodynamique et de Séismologie, Luxembourg, 99 pp, 1998

Hancılar U, Tüzün C, Yenidoğan C, Erdik M (2010) ELER software – a new tool for urban earthquake loss assessment. Nat Hazards Earth Syst Sci 10:2677–2696

HAZUS (1999) HAZUS earthquake loss estimation methodology. Technical Manual, Prepared by the National Institute of Building Sciences for the Federal Emergency Management Agency, Washington DC

Inoue T, Cornell CA (1990) Seismic hazard analysis of multi-degree-of-freedom structures. Reliab Marine Struct RMS-8, Stanford, CA, 70 pp

Jaiswal K, Wald D, Porter K (2010) A global building inventory for earthquake and risk management. Earthq Spect 26(3):731–748

Jayaram N, Baker JW (2009) Correlation model for spatially-distributed ground-motion intensities. Earthq Eng Struct Dyn

Kale Ö, Akkar S, Ansari A, Hamzehloo HA (2015) Ground-motion predictive model for Iran and Turkey for horizontal PGA, PGV, and 5% damped response spectrum: Investigation of possible regional effects. Bull Seism Soc Am 105(2A):963–980, Apr. https://doi.org/10.1785/0120140134

Kircher CA, Whitman RV, Holmes WT (2006) earthquake loss estimation methods. Nat Hazards Review 7(2):45–59

Krawinkler H (2002) A general approach to seismic performance assessment. In: Proceedings of the international conference on advances and new challenges in earthquake engineering research, ICANCEER, Hong Kong Vol, 173–180

Lagomarsino S, Giovinazzi S (2006) Macroseismic and mechanical models for the vulnerability and damage assessment of current buildings. Bull Earthq Eng 4(4)

Lu, DG, XH Yu, MM Jia (2012) Analytical formulations of fragility functions with applications to probabilistic , 15 WCEE, Lisboa

Luco N, Ellingwood BR, Hamburger RO, Hooper JD, Kimball JK, Kircher CA (2007) Risk-targeted versus current seismic design maps for the conterminous United States. In: SEAOC convention 2007 proceedings

Luco, N. and E. Karaca (2006), Improving Input into & Other Tools, /MAEC Workshop in Memphis, TN, October 11, 2006

McGuire (2004) Seismic hazard and risk analysis, monograph MNO-10, Earthquake Engineering Research Institute, Oakland, U.S.A

Melchers RE (1999) Structural and prediction. Wiley, New York

Moehle JP (2003) A framework for performance-based earthquake engineering. In: Proc. ATC-15-9 Workshop on the Improvement of Building Structural Design and Construction Practices, Maui, HI, June

Munich R (2016) Loss events worldwide 1980–2015, 10 costliest events ordered by overall losses (as at March 2016), Munich Reinsurance Company

Mouroux P, Le Brun B (2006) Presentation of RISK-UE Project. Bull Earthq Eng 4:323–339

Musson RMW (2000) The use of Monte Carlo simulations for seismic hazard assessment in the UK. Ann Geofis 43:1–9

OYO (2009) Updating Estimations of the Probable Earthquake in Istanbul - Final Report prepared for Istanbul Municipality, OYO International Corporation

Park J, Bazzurro P, Baker JW (2007) Modeling spatial correlation of ground motion intensity measures for regional seismic hazard and portfolio loss estimations. In: Takada Furuta (ed) Kanda.

Applications of statistics and probability in civil engineering. Taylor & Francis Group, London, pp 1–8

Pitilakis K, Crowley H, Kaynia A (eds) (2014a) SYNER-G: Typology definition and fragility functions for physical elements at seismic risk, vol 27, Geotechnical, geological and earthquake engineering. Springer, Heidelberg. ISBN 978-94-007-7872-6

Pitilakis K, Franchin P, Khazai B, Wenzel H (eds) (2014b) SYNER-G: systemic seismic vulnerability and risk assessment of complex urban, utility, lifeline systems and critical facilities. Methodology and applications, Geotechnical, geological and earthquake engineering. Springer, Heidelberg. ISBN 978-94-017-8834-2

Priestley MJN (2003) Myths and fallacies in earthquake engineering, revisited, The Mallet Milne Lecture. IUSS Press, Pavia, Italy

RISK-UE (2004) The European Risk-UE Project: An advanced approach to earthquake risk scenarios. (2001–2004) www.risk-ue.net

Ruiz-Garcia J, Miranda E (2007) Probabilistic estimation of máximum inelastic displacement demands for performance-based design. Earthq Eng Struct Dyn 36:1235–1254

Scherbaum F, Schmedes J, Cotton F (2004) On the conversion of source-to-site distance measures for extended earthquake source models. Bull Seism Soc Am 94:1053–1069

Silva V, Crowley H, Pagani M, Pinho R, Monelli D (2012) Development and application of openquake, an open source software for seismic risk assessment. In: Proc. 15 WCEE, Lisboa

Silva V, Crowley H, Pagani M, Monelli D, Pinho R (2013) Development of the OpenQuake engine, the Global Earthquake Model's open-source software for seismic risk assessment. Nat Hazards. https://doi.org/10.1007/s11069-013-0618-x

Spence R (Ed.), Erdik M (Rev.) (2007) Earthquake disaster scenario prediction and loss modelling for urban areas, LESSLOSS Report No. 2007/07, IUSS Press, Pavia, Italy

Strasser FO, Stafford PJ, Bommer JJ, Erdik M (2008) State-of-the-art of European earthquake loss estimation software. In: Proc. the 14 th World Conference on Earthquake Engineering October 12–17, 2008, Beijing, China

Stafford PJ, Strasser FO, Bommer JJ (2007) Preliminary report on the evaluation of existing methodologies, Report prepared for EU FP6 NERIES Project, Department of Civil & Environmental Engineering, Imperial College, London

Tyagunov S, Grunthal G, Wahlstrom R, Stempniewski L, Zschau J (2006) Seismic risk mapping for Germany. Nat Hazards Earth Syst Sci 6:573–586, 2006

Wagener T, Goda K, Erdik M, Daniell J, Wenzel F (2016) A spatial correlation model of peak ground acceleration and response spectra based on data of the Istanbul earthquake rapid response and early warning system. Soil Dyn Earthq Eng 85:166–178

Wald DJ, Jaiswal KS, Marano KD, Bausch DB, Hearne MG (2010) PAGER—Rapid assessment of an earthquake's impact. U.S. Geological Survey Fact Sheet 2010–3036, 4 p

Wald DJ, Quitoriano V, Heaton TH, Kanamori H (1999) Relationships between peak ground acceleration, peak ground velocity, and modified mercalli intensity in California. Earthq Spect 15(3):557–564

Wald DJ, Worden CB, Quitoriano V, Pankow KL (2006) ShakeMap® Manual, technical manual, users guide, and software guide,

Wang M, Takada T (2005) Macrospatial correlation model of seismic ground motions. Earthq Spect 21(4):1137–1156

Weatherill G, Silva V, Crowley H, Bazzurro P (2013) Exploring strategies for portfolio analysis in probabilistic seismic loss estimation. In: Proceedings of Vienna Congress on recent advances in earthquake engineering and structural dynamics (VEESD 2013), paper no. 303, Vienna, Austria

Wen YK, Ellingwood BR, Bracci J (2004) Vulnerability function framework for consequence-based engineering, Mid-America Earthquake Center

Whitman RV, Reed JW, ST Hong (1973) Earthquake Damage Probability Matrices. In: Proceedings of the 5th World Conference on Earthquake Engineering, Rome, Italy

Whitman RV, Anagnos T, Kircher CA, Lagorio HJ, Lawson RS, Schneider P (1997) Development of a national earthquake methodology. Earthq Spect 13(4):643–661

Wong FS, Chen H, Dong W (2000) Uncertainty modeling for disaster loss estimation. In: Proc. 12 WCEE, New Zealand

World Housing Encyclopedia online database http://www.world-housing.net

Wu J, Baker JW (2014) Ground motion modeling for risk and reliability assessment of San Francisco infrastructure systems. In: Proceedings of the tenth U.S. national conference on earthquake engineering, Anchorage, Alaska

Yepes-Estrada C, Silva V, Crowley H (2014) GEM vulnerability database for the openquake-platform. In: Proc., 2nd European conference on earthquake engineering and seismology, Istanbul, Aug 25–29, 2014

Yepes-Estrada C, Silva V, Rossetto T, D'Ayala D, Ioannou I, Meslem A, Crowley H (2016) The global earthquake model physical vulnerability database. Earthq Spect 32(4):2567–2585

Yücemen MS (2013) Probabilistic assessment of earthquake insurance rates, handbook of and management of civil infrastructure systems. In: Tefamariam S, Goda K (eds) Chapter 29. Woodhead Publishing Ltd., pp 787–814

Chapter 7
European Exposure and Vulnerability Models: State-of-The-Practice, Challenges and Future Directions

H. Crowley

7.1 Introduction

Initiated as a Joint Research Activity of the European Commission's Horizon 2020 Project SERA (www.sera-eu.org), a European Seismic Risk Model (ESRM20) (Crowley et al. 2019) is being developed using open/publicly available data on all components of seismic risk from catalogues, to active faults, building data and vulnerability models. This model will be released by the European Facilities for Earthquake Hazard and Risk (EFEHR) Consortium (www.efehr.org) under the following general principles for open, transparent and reproducible hazard and risk models:

- Reproducibility. Reproducibility of an experiment—or of a complex model—is one basic and unavoidable principle of modern science (Popper 2002).
- Transparency. Transparency ensures that all aspects of scientific methods and results are available for critique, checking, compliment, or reuse.
- FAIR principles (European Commission 2016). The data and models used or produced should be Findable, Accessible, Interoperable and Reusable.
- Respect of the intellectual property and clear scientific ownership. A proper recognition (and citation) of data support or scientific contribution is indispensable.

An update to the 2013 European seismic hazard model (ESHM13: Woessner et al. 2015) together with a regional site amplification model (based on the methodology presented in Weatherill et al. 2020) will provide the probabilistic estimates of surface ground shaking for this risk model. This chapter summarises the current status of the exposure and vulnerability components of this seismic risk model, addresses where the key modelling challenges presently lie, and looks towards the future directions

H. Crowley (✉)
European Centre for Training and Research in Earthquake Engineering (EUCENTRE), Via Ferrata 1, 27100 Pavia, Italy
e-mail: helen.crowley@eucentre.it

© The Author(s) 2021
S. Akkar et al. (eds.), *Advances in Assessment and Modeling of Earthquake Loss*,
Springer Tracts in Civil Engineering,
https://doi.org/10.1007/978-3-030-68813-4_7

that are being explored to address those shortcomings and move towards improved European seismic risk and loss modelling under the general principles outlined above.

7.2 Exposure Modelling

7.2.1 Summary of European Exposure Model

A European exposure model describing the spatial distribution of residential, commercial and light industrial buildings in terms of building count, population, and replacement cost, and classified in terms of building classes, is being developed for 44 European countries (Crowley et al. 2020a).

These residential and non-residential exposure models have been derived based on the latest national population and dwelling censuses, socio-economic indicators (e.g. labour force, population and floor area per worker per economic sector), mapping schemes (to map the available data to building classes) developed together with local experts, as well as engineering judgment. All of the source data that has been collected, as well as the assumptions used in the development of each version of the model, are being openly released on a GitLab repository[1] with a Creative Commons license. This repository will also be used to store the final exposure models for all European countries, which will be released towards the end of 2020.

The European exposure model contains a total of around 145 million buildings with a total replacement cost (of structural, non-structural elements and contents) of around 45–50 Trillion EUR, of which 20% is attributed to industrial buildings, 20% to commercial buildings and 60% to residential buildings. The top 10 countries in terms of number and value of buildings is shown in Fig. 7.1.

Around 70% of the total buildings in Europe are found in these top 10 countries, whereas about 80% of the value is concentrated in 10 countries. Poland, Turkey and Romania, which have a large number of buildings and are found in the first figure are replaced by the Netherlands, Sweden and Switzerland in the second figure because, despite having a lower number of buildings, they have a higher total replacement value due to the much higher construction costs in these countries. It has also been found that around 35% of the European population is exposed to moderate levels of seismic hazard (>0.1 g) (Crowley et al. 2020a).

Maps of the exposure models, and associated web services following Open GeoSpatial Consortium (OGC) standards, are being made available through a web platform (see https://maps.eu-risk.eucentre.it/ and https://eu-risk.eucentre.it/web-ser vices/) thus allowing the exposure data to be easily integrated within other web applications and platforms. Figure 7.2 presents one of these maps which shows the the distribution of total replacement cost on a hexagonal grid, with a spacing of

---

[1] https://gitlab.seismo.ethz.ch/efehr/esrm20_exposure.

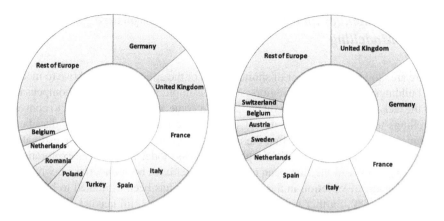

Fig. 7.1 Top 10 European countries in terms of number of buildings (left) and replacement cost (right)

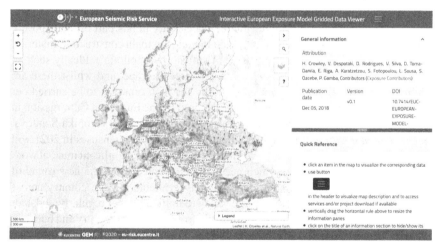

Fig. 7.2 Screenshot of the gridded exposure model viewer showing the distribution of total replacement cost on a hexagonal grid (https://maps.eu-risk.eucentre.it/map/european-exposure-gridded-data)

0.30×0.34 decimal degrees (approximately 1000 km^2 at the equator) (using the methodology described in Silva et al. 2020).

7.2.2 Challenges and Future Directions in Exposure Modelling

There are, however, a number of shortcomings of the approach used above to model the buildings at risk over large regions. Many assumptions are required to compensate for the lack of open/public data on buildings (e.g. the assumptions needed to convert dwellings to buildings, or the use of labour force statistics to spatially distribute commercial buildings), and often the model uncertainty in not explicitly estimated or documented, nor propagated through the risk/loss model. Some initial explorations of the uncertainty in the European exposure model have been undertaken, whereby the coefficient of variation in the replacement cost has been estimated to be of the order of 40–50%. Further sensitivity studies are however still needed, in particular related to the impact on the distribution of the building classes, which is often based on expert judgment.

Some countries in Europe (Italy, Portugal and Greece) have undertaken a building census in conjunction with the national population and dwelling census, and they have classified the buildings already into classes that correlate with the seismic performance of buildings, thus reducing the uncertainty in this part of the exposure model. The main attributes that are collected include the main construction material, total number of storeys, age and presence of soft storeys ('pilotis'). Ideally such an effort would be carried out in more countries across Europe, and whilst there are ongoing efforts within some countries to lobby for such censuses to be carried out as input to the National Risk Assessments, required by the European Commission in support of the Sendai Framework for Disaster Risk Reduction (Veronika Sendova, *personal communication*), it is unlikely that the next round of censuses in 2021 will differ significantly from those undertaken in 2011. Given the significant manual work used to develop these models (which needs to be repeated when the new round of census data will be collected and made publicly available in each country across Europe), the resulting models are "static" and are unlikely to get regularly updated.

The resolution of the data varies significantly from country to country and between residential, industrial and commercial buildings. For most countries the distribution of buildings in the industrial exposure model is based on the 30 arcsec (approximately 1 km at the equator) grid of industrial built-up area developed by Sousa et al. (2017) and is thus available at a very high resolution. The residential and commercial models, on the other hand, depend on the resolution at which the census data for dwellings, buildings or labour force statistics is publicly available. Figure 7.3 shows the variation in the resolutions (in terms of highest administrative boundaries) for all countries, which shows how much the resolution varies and highlights that the commercial resolution is quite poor in most European countries.

As commonly known, the uncertainty in the location of assets introduces a bias in the level of ground shaking and, consequently, the level of damage (see e.g. Bal et al. 2010). Moreover, the bias can be magnified by the various site conditions at different locations. A study to investigate the impact of the spatial resolution of the exposure on the risk metrics being developed for the European Seismic Risk Model has been

Fig. 7.3 Highest administrative level resolution of the exposure models for residential (left) and commercial (right) buildings

initiated (see Crowley et al. 2020b). The residential and commercial occupancies have been disaggregated to six resolutions 30, 60, 120, 240, 480 and 960 arcsec. In this process, buildings are redistributed using remote sensing information at 38 × 38 squared metre resolution and then aggregated to the different grid resolutions. More details on the disaggregation methodology can be found in Dabbeek and Silva (2020). In addition to the gridded exposure models, three additional workflows (wf) were investigated: (1) locations based on the centroid of administrative unit and the closest site conditions, (2) locations based on the centroid of administrative unit and average site conditions weighted by the density of built-up areas across the unit, (3) locations based on the maximum density of built-up areas and the average site conditions weighted by the density of built-up areas across the unit.

The risk metrics have been calculated using the probabilistic event-based calculator of the OpenQuake-engine (Pagani et al. 2014; Silva et al. 2014) with stochastic event sets covering 100,000 years for all modelling cases. Figure 7.4 presents the change in the national average annual loss (AAL) between the different exposure modelling cases and the benchmark model (30 arcsec). For the gridded exposure, the results indicate relatively stable losses up 240 resolution with a maximum difference of 3%. After this, the results become inaccurate, reaching a maximum difference of 25%, which can be seen in the case of Iceland and Turkey. Similar analyses were undertaken for the risk metrics at the sub-national level (aggregated to the first administrative level). These analyses illustrated that at the national level, the AAL is better estimated with the second workflow described above (wf2). The percentage change is likely to be proportional to the size of the administrative boundary, population distribution and the attenuation of ground motions and plots to demonstrate this will be produced in future versions of the study. These relationships could then be used to help identify the lowest resolution that could be used in regional/national exposure modelling in a given country to ensure a balance between computational efficiency and accuracy.

To address some of the limitations described above, the future of exposure modelling is likely to focus on producing dynamic high-resolution exposure models with the necessary tools and web services that will allow them to be automatically

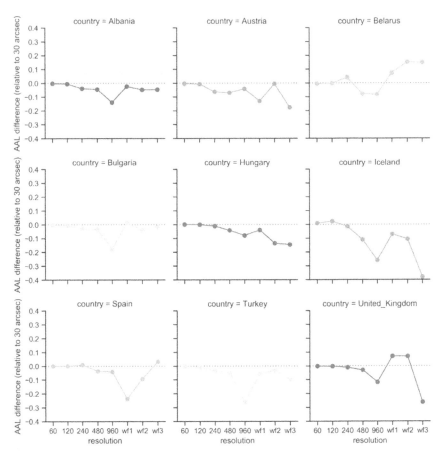

Fig. 7.4 Change in national average annual loss (AAL) due to changes in exposure resolution for nine European countries with low exposure resolution. Results are provided relative to the benchmark case (30 arc-second exposure)

updated. Within the European Horizon 2020 RISE project (www.rise-eu.org), an effort led by GFZ Potsdam is being undertaken to develop a high-resolution Global Dynamic

Exposure (GDE) model. The GDE aims to describe exposure on the building level of every building on Earth employing a fully open big-data approach including open geographic data such as OpenStreetMap,[2] open remote-sensing data, machine learning, and other open data like cadastral data-services. The GDE provides a server infrastructure to automatically compute exposure indicators for ~375 million buildings at a global scale (a number which is growing by approx. 150,000 buildings daily as more buildings are mapped in OpenStreetMap). Some of these indicators are

---

[2]https://www.openstreetmap.org.

shown on the OpenBuildingMap[3] and its 3D version.[4] Currently, the high-resolution building data from GED is being combined with the building classifications from the European exposure models described above as a first step to producing a high-resolution European exposure model that can be used for earthquake loss assessment under specific scenario events.

7.3 Vulnerability Modelling

7.3.1 Summary of European Vulnerability Model

Whilst vulnerability models can be developed directly from empirical loss data (e.g. Jaiswal et al. 2009), often the resolution and quality of ground motion and loss data in public databases is not sufficient for this purpose, and vulnerability models are thus commonly developed by combining fragility functions with consequence models, which define the probability of loss, conditional on the level of damage.

Fragility models for the elements at risk within an exposure model provide the probability of reaching or exceeding a set of damage states, conditional on the level of ground shaking. Whilst these models can be developed using observed damage data, the large uncertainties in the ground shaking to which the buildings have been subjected often mean that the resulting functions are flatter and highly uncertain (e.g. Ioannou et al. 2014). Analytical modelling is thus preferred as hazard consistent ground shaking at the site can be considered, the relative difference between building classes (some of which may not yet have experienced earthquake damage in past events) can be explicitly modelled, and data on the characteristics of specific buildings (when available) can be used to update the models. The latest developments, as well as limitations, in analytical vulnerability modelling has been covered in Silva et al. (2019).

A European vulnerability database, comprising capacity curves, fragility functions, damage-loss models and vulnerability functions has been compiled within the SERA project and is available on a GitLab repository (Romão et al. 2020). This database currently has 828 models from 63 separate studies obtained from the literature. Such a database is particularly useful for sanity checking new fragility models as it allows modellers to compare their models with those from the literature (see Crowley et al. 2020b).

In addition to collecting existing vulnerability models, a new set of models for the building classes in the European Seismic Risk Model is being developed. As part of this effort, the spatial and temporal evolution of design codes for reinforced concrete buildings across Europe has been studied (Crowley et al. 2021) and the basic principles of seismic design according to four main categories of design (pre-code—CDN,

---

[3] https://www.openbuildingmap.org.
[4] https://obm3d.gfz-potsdam.de.

low—CDL, moderate—CDM and high—CDH) has been used to design prototype buildings, which have then been numerically modelled to obtain their lateral strength and deformation capacity. Buildings of design class CDN were typically designed to older codes (from before the 1960's) that used allowable stresses and very low material strength values and considered predominantly the gravity loads. Buildings of design class CDL were designed considering the seismic action by enforcing values of the design lateral force coefficient (defined as the lateral force applied as a fraction of the weight of the building). Structural design for these codes was typically based on material-specific standards that used allowable stress design or a stress-block approach. Seismic design including modern concepts of ultimate capacity and partial safety factors (limit state design) was the basis of the CDM category of codes. The seismic action was also accounted for in the design by enforcing values for the lateral force coefficient. Finally, the CDH class refers to modern seismic design principles that account for capacity design and local ductility measures, similar to those available in Eurocode 8 (CEN 2004).

Numerical models of the MDOF designed buildings are produced and pushover curves are obtained in the two orthogonal directions and these are transformed into SDOF systems (or, for building classes which are not explicitly designed and numerically modelled, the SDOF systems are directly inferred following the approach of Martins and Silva 2020) and these SDOF models are then subjected to a range of ground motion recordings (through dynamic nonlinear analysis) to model the relationship between ground shaking intensity and displacement response, using the censored cloud approach described in Crowley et al. (2017) and Martins and Silva (2020). Uncertainties in the characteristics of the buildings (geometrical and material), the design parameters, the quality of construction (and thus adherence to code), the displacement thresholds to damage, and the record to record variability can all be accounted for in the procedure and are modelled as aleatory variabilities represented in the final dispersion of the fragility functions, constructed from the cloud as illustrated in Fig. 7.5.

The fragility functions are then converted into vulnerability models using damage-loss models which provide loss ratios for each damage state (slight, moderate,

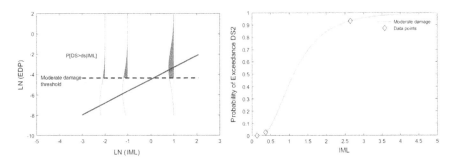

Fig. 7.5 Construction of a fragility function following the cloud analysis approach (Martins and Silva 2020)

extensive and complete). For losses due to the repair of damage, the loss ratios are inferred from a number of existing studies (e.g. Di Pasquale and Goretti 2001; FEMA 2004; Kappos et al. 2006; Bal et al. 2008). For loss of life, the probability of collapse given complete damage is first estimated by combining the proposals from FEMA (2004) with engineering judgment, and comparing these with observed damage data available in databases such as the Italian Department of Civil Protection's Da.D.O. database (egeos.eucentre.it/danno_osservato/web/danno_osservato, Dolce et al. 2019), and the Cambridge Earthquake Impact Database (https://www.ceqid.org). Fatality ratios (i.e. the probability of loss of life given collapse for different building classes) are still being developed through the evaluation of fatality data from a number of past damaging earthquakes.

7.3.2 Challenges and Future Directions in Vulnerability Modelling

Whilst the latest approaches for vulnerability modelling account for a wide range of uncertainties, modelling uncertainty is typically not considered, where the latter is defined herein as the uncertainty associated with the selected modelling approach, rather than the commonly considered 'parameter uncertainty' which is the uncertainty associated with the parameters of a particular modelling approach. As more experimental tests of components and full-scale buildings become available, there is scope to quantify the bias or lack of precision of the structural modelling methodology used in the development of analytical fragility functions (e.g. Bradley 2013). However, even when the selected modelling approach is tested/calibrated against some experimental tests, blind prediction exercises show that results from plausible models can still vary significantly (see e.g. Terzic et al. 2015). The buildings for which the fragility functions are the being developed will not necessarily have the same characteristics as those for which the numerical modelling approach has been calibrated against. Sensible variations of the model should thus still be undertaken when developing fragility functions for a given structural typology, or the impact should be applied *ex post* through engineering judgment, based on the normalised results of other similar studies into modelling uncertainty – of which more are needed. An example publication that proposes values for modelling uncertainty is FEMA P-58 (FEMA 2018). Such variations in the model might include, for reinforced concrete buildings, varying the assumptions regarding rigidity of the beam-column joints, the bond between rebars and concrete, the contact between infill panels and the frame, and for masonry buildings the interlock between orthogonal walls, connection between slabs, roof and walls or assumptions on the equivalent frame discretisation. The modelling of this epistemic uncertainty should thus become standard practice in future analytical fragility modelling, and might be based on a backbone approach with the aleatory model uncertainty represented through a logic tree (Crowley et al. 2017), an approach that is being increasingly used for ground motion modelling (e.g.

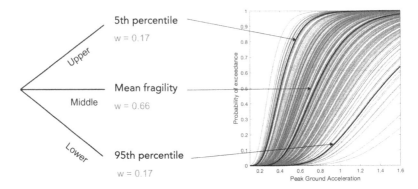

Fig. 7.6 Illustrative example of the representation of the model uncertainty in fragility functions as a three-point distribution used for the logic tree

Atkinson et al. 2014; Douglas 2018). Figure 7.6 shows an example of how the epistemic uncertainty in the fragility models (by producing a large number of fragility functions due to the variations in the modelling approach) can be transformed to a three-point distribution that is used for the logic tree. The advantage of including the model uncertainty as an epistemic uncertainty, rather than an aleatory variability, allows the correlation of this uncertainty across building classes that are based on similar modelling approaches to be more readily accounted for in the risk model.

In order to improve the transparency and reproducibility of fragility models and to render more explicit the uncertainties that have been modelled, it is recommended that, in addition to providing the parameters of the models through vulnerability databases such as the one described above, the underlying data (e.g. SDOF model parameters, selected records, damage thresholds) and the software used to develop the models should also be made openly available. The Global Earthquake Model (GEM) is currently developing open source Python scripts and tools (the 'vulnerability modeller's toolkit') that follow the vulnerability methodology used by GEM in their Global Seismic Risk Model (Silva et al. 2020; Martins and Silva 2020). These tools will allow users to produce fragility models that are based on a common methodology and can be readily compared, and advanced users will be able to make modifications to the scripts that can be openly shared.

Another effort that is being undertaken to improve the reliability of future vulnerability modelling is the formalisation of a testing framework for risk models (Crowley et al. 2020b, c). Simple sanity checks, so-called 'unit tests' can be included in software for developing fragility functions (such as the one described above) to ensure the median and dispersion values are within sensible ranges, and to compare with existing functions from the literature. However, it should be considered when undertaking such comparisons that many of models from the academic literature have not been calibrated or tested using past earthquake damage and loss data. Hence, although comparisons with existing models is an important test, it is even more important to ensure that the proposed models are tested against empirical data. Useful, and openly

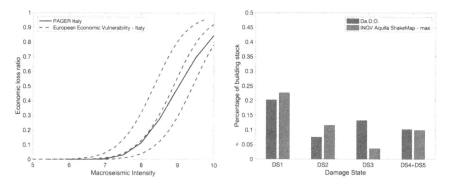

Fig. 7.7 Example tests of the European vulnerability model: comparison with the PAGER vulnerability model (left), comparison of estimated and observed damage for the L'Aquila earthquake using damage data from the Da.D.O. database (https://egeos.eucentre.it/danno_osservato/web/danno_osservato) (right)

available, data for this purpose includes the empirical vulnerability models developed by PAGER (Jaiswal et al. 2009; Jaiswal and Wald 2013), as well as fatality, economic loss and damage data from various databases including the Centre for Research on the Epidemiology of Disasters (CRED)'s EMDAT database (EMDAT 2019), the Italian Department of Civil Protection's Da.D.O. database (Dolce et al. 2019), NOAA's Significant Earthquake Database (NGDC/WGS), and the Cambridge Earthquake Impact Database (www.ceqid.org). Despite the current availability of damage and loss data for the verification of seismic risk models, continued efforts to standardise and harmonise the collection of open and publicly available consequence data is still needed. Efforts to combine these data sources with the USGS ShakeMaps for all earthquakes in Europe above magnitude 4 since 1960 are currently being undertaken by the author to produce an open standardised data source that can be used for the testing of European risk models.

Figure 7.7 shows two examples of tests that have been undertaken with the European vulnerability model (Crowley et al. 2020b). In the first example, a mean vulnerability function calculated through an exposure-weighted combination of all the building classes in the country has been produced and compared with the empirical models developed by PAGER, following conversion of the spectral ordinates to macroseismic intensity (with the associated uncertainty in the conversion shown by the mean and +/1 standard deviation vulnerability curves in Fig. 7.7). In the second example, the INGV ShakeMap for the 2009 L'Aquila earthquake has been used together with the exposure and vulnerability models to estimate the damage distribution and this has been compared with the damage reported in the Da.D.O. database (using the method outlined in Silva and Horspool 2019).

7.4 Concluding Remarks

This chapter has presented the latest status of the exposure and vulnerability components of the European Seismic Risk Model (ESRM20) which is under development
and will be released in autumn 2020 by the risk services of the European Facilities
for Earthquake Hazard and Risk (EFEHR) Consortium.[5] These models follow the
state-of-the-practice of large-scale, regional exposure and vulnerability modelling.
Some of the challenges in the current practice, such as limited access to public data,
manual updating, difficulties in reproducing current models, and lack of testing, have
been discussed herein and the future directions being taken to address these issues
have been outlined. On the whole, it is believed that a move towards releasing all
underlying data sources of the components of risk models in an open and transparent
manner, together with the software used to develop them, will ensure the continued
improvement of European risk modelling.

References

Atkinson GM, Bommer JJ, Abrahamson NA (2014) Alternative approaches to modeling epis-
    temic uncertainty in ground motions in probabilistic seismic-hazard analysis. Seismo Res Lett
    85(6):1141–1144. https://doi.org/10.1785/0220140120
Bal IE, Crowley H, Pinho R (2008) Detailed assessment of structural characteristics of Turkish RC
    building stock for loss assessment models. Soil Dyn Earthq Eng 28:914–932
Bal IE, Bommer JJ, Stafford PJ, Crowley H, Pinho R (2010) The influence of geographical resolution
    of urban exposure data in an earthquake loss model for Istanbul. Earthq Spect 26(3):619–634.
    https://doi.org/10.1193/1.3459127
Bradley B (2013) A critical examination of seismic response uncertainty analysis in earthquake
    engineering. Earthq Eng Struct Dyn 42(11):1717–1729
CEN (2004) Eurocode 8: design of structures for earthquake resistance. European Standard,
    European Committee for Standardisation, Brussels
Crowley H, Polidoro B, Pinho R, Van EJ (2017) Framework for developing fragility and consequence
    models for local personal risk. Earthq Spect 33(4):1325–1345
Crowley H, Rodrigues D, Silva V, Despotaki V, Marins L, Romão X, Castro JM, Pereira N, Pomonis
    A, Lemoine A, Roullé A, Tourlière B, Weatherill G, Pitilakis K, Danciu L, Correira AA, Akkar
    S, Hancilar U, Covi P (2019) The European seismic risk model 2020 (ESRM20). In: Proceedings
    of 2nd international conference on natural hazards and infrastructure, ICONHIC 2019. Chania,
    Crete
Crowley H, Despotaki V, Rodrigues D, Silva V, Toma-Danila D, Riga E, Karatzetzou A, Fotopoulou
    S, Zugic Z, Sousa L, Ozcebe S, Gamba P (2020a) Exposure model for European seismic risk
    assessment. Earthq Spect. https://doi.org/10.1177/8755293020919429
Crowley H, Dabbeek J, De Maio FV, Despotaki V, Rodrigues D, Faravelli M, Borzi B, Silva V,
    Martins L, Kalakonas P, Weatherill G, Riga E, Karatzetzou A, Pitilakis K, Anastasiadis A, Pitilakis
    D, Fotopoulou S, Michelini A, Faenza L (2020b) D26.8 testing and verification of the European
    Seismic Risk Model (ESRM20), SERA Project Deliverable. www.sera-eu.org

---

[5]https://eu-risk.eucentre.it/.

Crowley H, Despotaki V, Silva V, Dabbeek J, Romão X, Daniell J, Veliu E, Bilgin H, Adam C, Deyanova M, Ademović N., Atalic J, Riga E, Karatzetzou A, Bessason B, Sendova V, Toma-Danila D, Zugic Z, Akkar S, Hancilar U (2021) Model of seismic design lateral force levels for the existing European building stock, Bull Earthq Eng. https://doi.org/10.1007/s10518-021-010 83-3

Crowley H, Silva V, Kalakonas P, Martins L, Weatherill G, Pitilakis K, Riga E, Borzi B, Faravelli M (2020c) Verification of the European seismic risk model (ESRM20). In: Proceedings of 17th world conference on earthquake engineering. Sendai, Japan

Dabbeek J, Silva V (2020) Modeling the residential building stock in the Middle East for multi-hazard risk assessment. Nat Hazards 100:781–810. https://doi.org/10.1007/s11069-019-03842-7

Di Pasquale G, Goretti A (2001) Vulnerabilità funzionale ed economica degli edifici residenziali colpiti dai recenti eventi sismici italiani. In: Proceedings of the 10th National Conference "L'ingegneria Sismica in Italia". Potenza-Matera, Italy

Dolce M, Speranza E, Giordano F, Borzi B, Bocchi F, Conte C, Di Meo A, Faravelli M, Pascale V (2019) Observed damage database of past Italian earthquakes: the Da.D.O WebGIS. Bollettino di Geofisica Teorica ed Applicata 60(2):141–164. https://doi.org/10.4430/bgta0254

Douglas J (2018) Capturing geographically-varying uncertainty in earthquake ground motion models or what we think we know may change. In: Pitilakis K (ed) Recent advances in earthquake engineering in Europe, Chapter 6, pp 153–181

EMDAT (2019) International disasters database of the centre for research on the epidemiology of disasters. https://www.emdat.be/

European Commission (2016) Guidelines on FAIR data management in horizon 2020, July 2016. https://ec.europa.eu/research/participants/data/ref/h2020/grants_manual/hi/oa_pilot/h2020-hi-oa-data-mgt_en.pdf

FEMA (2004) HAZUS-MH technical manual. Federal Emergency Management Agency, Washington DC, USA

FEMA (2018) Seismic performance assessment of buildings. Volume 1 – Methodology. FEMA P-58-1, Second Edition, Federal Emergency Management Agency, Washington DC

Ioannou I, Douglas J, Rossetto T (2014) Assessing the impact of ground-motion variability and uncertainty on empirical fragility curves. Soil Dyn Earthq Eng. https://doi.org/10.1016/j.soildyn.2014.10.024

Jaiswal K, Wald D (2013) Estimating economic losses from earthquakes using an empirical approach. Earthq Spectra 29(1):309–324

Jaiswal K, Wald D, Hearne M (2009) Estimating casualties for large worldwide earthquakes using an empirical approach, US Geological Survey Open-File Report 1136

Kappos A, Panagopoulos G, Panagiotopoulos C, Penelis G (2006) A hybrid method for the vulnerability assessment of R/C and URM buildings. Bull Earthq Eng 4(4):391–413. https://doi.org/10.1007/s10518-006-9023-0

Martins L, Silva V (2020) Development of a fragility and vulnerability model for global seismic risk analyses. Bull Earthq Eng. https://doi.org/10.1007/s10518-020-00885-1

National Geophysical Data Center/World Data Service (NGDC/WDS): significant earthquake database. National Geophysical Data Center, NOAA. https://doi.org/10.7289/V5TD9V7K. Access date: 22 Apr 2020

Pagani M, Monelli D, Weatherill G, Danciu L, Crowley H, Silva V, Henshaw P, Butler L, Nastasi M, Panzeri L, Simionato M, Vigano D (2014) OpenQuake Engine: an open hazard (and risk) software for the global earthquake model. Seism Res Lett 85(3):692–702

Popper K (2002) The logic of scientific discovery. Rutledge classic, p 545

Romão X, Pereira N, Castro JM, De Maio F, Crowley H, Silva V, Martins L (2020) European Building Vulnerability Data Repository (Version v1.1) [Data set]. Zenodo. http://doi.org/10.5281/zenodo.4087810

Silva V, Crowley H, Pagani M, Monelli D, Pinho R (2014) Development of the OpenQuake engine, the global earthquake model's open-source software for seismic risk assessment. Nat Hazards. https://doi.org/10.1007/s11069-013-0618-x

Silva V, Akkar S, Baker J, Bazzurro P, Castro JM, Crowley H, Dolsek M, Galasso C, Lagomarsino S, Monteiro R, Perrone D, Pitilakis K, Vamvatsikos D (2019) Current challenges and future trends in analytical fragility and vulnerability modelling. Earthq Spect 35(4):1927–1952

Silva V, Amo-Oduro D, Calderon A, Costa C, Dabbeek J, Despotaki V, Martins L, Pagani M, Rao A, Simionato M, Viganò D, Yepes-Strada C, Acevedo A, Crowley H, Horspool N, Jaiswal K, Journeay M, Pittore M (2020) Development of a global seismic risk model. Earthq Spect. https://doi.org/10.1177/8755293019899953

Silva V, Horspool N (2019) Combining USGS XE "USGS" shakemaps and the openquake-engine for damage and loss assessment. Earthq Eng Struct Dynam. https://doi.org/10.1002/eqe.3154

Sousa L, Silva V, Bazzurro P (2017) Using open-access data in the development of exposure data sets of industrial buildings for earthquake risk modeling. Earthq Spect 33(1):63–84

Terzic V, Schoettler MJ, Restrepo JI, Mahin SA (2015) Concrete column blind prediction contest 2010: outcomes and observations. PEER Report 2015/01. Pacific Earthquake Engineering Research Center, Berkeley, CA

Weatherill G, Kotha SR, Cotton F (2020) Re-thinking site amplification in regional seismic risk assessment. Earthq Spect. https://doi.org/10.1177/8755293019899956

Woessner J, Danciu L, Giardini D, Crowley H, Cotton F, Grunthal G, Valensise G, Arvidsson R, Basili R, Demircioglu M, Hiemer S, Meletti C, Musson R, Rovida A, Sesetyan K, Stucchi M (2015) The 2013 European seismic hazard model—key components and results. Bull Earthq Eng 13(12):3553–3596. https://doi.org/10.1007/s10518-015-9795-1

Chapter 8
Risk Oriented Earthquake Hazard Assessment: Influence of Spatial Discretisation and Non-ergodic Ground-Motion Models

Peter J. Stafford

Abstract Three important aspects of ground-motion modelling for regional or portfolio risk analyses are discussed. The first issue is the treatment of discretisation of continuous ground-motion fields for generating spatially correlated discrete fields. Shortcomings of the present approach in which correlation models based upon point estimates of ground motions are used to represent correlations within and between spatial regions are highlighted. It is shown that risk results will be dependent upon the chosen spatial resolution if the effects of discretisation are not adequately treated. Two aspects of non-ergodic groundmotion modelling are then discussed. Correlation models generally used within risk modelling are traditionally based upon very simple partitioning of ground-motion residuals. As regional risk analyses move to non-ergodic applications where systematic site effects are considered, these correlation models (both inter-period and spatial models) need to be revised. The nature of these revisions are shown herein. Finally, evidence for significantly reduced between-event variability within earthquake sequences is presented. The ability to progressively constrain location and sequence-dependent systematic offsets from ergodic models as earthquake sequences develop can have significant implications for aftershock risk assessments.

8.1 Introduction

Seismic risk analyses have traditionally been built upon existing tools developed for the purposes of evaluating seismic hazard. These seismic hazard analyses are always conducted for a single spatial location and have traditionally made use of the ergodic assumption (Anderson and Brune 1999), with exceptions being limited to high-level applications for critical facilities such as nuclear power plants, e.g., Rodriguez-Marek et al. (2014).

P. J. Stafford (✉)
Imperial College London, London SW7 2AZ, UK
e-mail: p.stafford@imperial.ac.uk

© The Author(s) 2021
S. Akkar et al. (eds.), *Advances in Assessment and Modeling of Earthquake Loss*,
Springer Tracts in Civil Engineering,
https://doi.org/10.1007/978-3-030-68813-4_8

For regional, or large-portfolio, risk analyses, ground-motion demands need to be prescribed at multiple spatial locations simultaneously, and these spatial locations often represent broader spatial regions around those locations within the analysis framework. Issues associated with the discretisation of ground-motion fields and exposure distributions are often over-looked. In particular, ground motion fields are developed using statistical properties between individual points, rather than between spatial regions (Stafford 2012).

The ergodic assumption, in the context of ground-motion modelling, is the assumption that the statistical properties of ground-motions at one particular location can be represented by pooling data from many different spatial locations with nominally similar characteristics. This assumption is necessary because individual sites have insufficient numbers of ground-motion recordings to permit robust site-specific ground-motion models to be developed. As the data comes from nominally similar spatial locations, the actual differences from site-to-site and region-to-region that remain within the data has impacts upon both the median predictions of ground-motion models and the associated variability derived from the data.

The application of the ergodic assumption therefore enables large databases of empirical observations to be compiled, and for robust ground-motion models to then be derived. However, the associated cost is that the derived ergodic ground-motion model is calibrated to this ergodic database rather than to the target site, and to the most relevant rupture scenarios that drive the hazard and risk at this site. Recent efforts (Kuehn et al. 2016; Landwehr et al. 2016; Stafford 2014; Stafford 2019) have looked to develop ground-motion models that make use of ergodic databases, but still allow for site- or region-specific features to be accounted for within partially non-ergodic frameworks. An aspect of non-ergodic ground-motion modelling that has received limited attention thus far is the impact that relaxing the assumption has upon correlation models that are required within risk analyses.

The present chapter focusses upon aspects of these two issues: impacts of spatial discretisation upon correlation models; and, non-ergodic ground-motion modelling issues, with a particular focus upon spatial correlation and aftershock sequences. The following section, Sect. 8.2, discusses the impacts of discretisation upon correlations that are required within risk analyses. Thereafter, Sect. 8.3 discusses the impacts of non-ergodic ground-motion models upon spatial correlations. Section 8.4 then looks at how non-ergodic concepts can be used to refine aftershock risk analyses, before the chapter closes with some high-level conclusions.

8.2   Correlations Among Intensity Measures

Models that have been published to represent correlations among intensity measures fall into two broad classes: those that represent correlations between two different intensity measures at a single spatial location, e.g., Baker and Bradley (2017); Baker and Jayaram (2008), and those that represent the spatial location of two intensity measures (potentially the same intensity measure) at two different spatial locations, e.g., Foulser Piggott and Stafford (2012); Jayaram and Baker (2009). These models are all derived on the basis of point observations of intensity measure fields because recording instruments at themselves located at particular points in space.

However, within portfolio risk analyses it is not usually feasible to perform calculations for each structure within the portfolio. Rather, buildings are grouped into a set of structural classes that have different representative structural characteristics, and intensity measures are computed at distinct locations that actually represent discrete spatial regions. Ideally, the results of a risk analysis that one obtained from considering every building within the portfolio should be the same (or, on average, very similar) to that obtained from working with discrete building classes and spatially-discretised fields of intensity measures. The only way that this ideal scenario can be achieved is if a great deal of care is taken to ensure the appropriate mapping between correlations and covariances between points and those over spatial regions. Previous attempts to look at the influence of spatial discretisation upon risk results (Bal et al. 2010) have not appropriately dealt with the relation between point-to-point spatial correlations and region-to-region correlations.

The types of correlations that may need to be considered within a regional risk analysis are shown schematically in Fig. 8.1. In this figure, ground-motions are computed at the white nodal locations within each grey cell. These cells can contain multiple structures. The leftmost panel shows a case where we have buildings from the same class present within a single cell. Given that all of these buildings have the same fragility curves, requiring the same intensity measure as an input, and that this intensity measure is only predicted at a single location within the cell, the demands upon all buildings within the cell are treated as being identical. Clearly, that modelling representation is not consistent with reality, and the quality of the assumption degrades as the spatial resolution reduces.

To enable our risk results to scale appropriately for different spatial correlations we need to account for the spatial differences in building locations within a given cell.

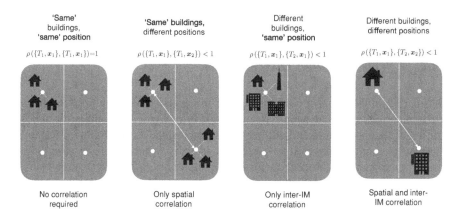

Fig. 8.1 Correlation cases to consider in portfolio risk analyses. Common assumptions about the correlations made in each case are annotated above and below each panel

This is true for all cases shown in Fig. 8.1 and influences the effective correlations that we use for buildings of the same class, and of different classes. When also considering spatial correlations across different cells we also need to account for the different site-to-site distances that can arise across those two cells.

To explain these issues more formally, the next section introduces how correlations between two points are traditionally handled, and then explains what the impact of spatial discretisation is for these models.

8.2.1 Point-Wise Correlations

Figure 8.1 showed that we need to have general correlation models that describe correlations between two buildings, characterised by response periods T_i and T_j[1], and located at sites $\mathbf{x}_p$ and $\mathbf{x}_q$, respectively. That is, we need to define the correlation between the intensity measures $\ln Sa(\{T_i, \mathbf{x}_p\})$ and $\ln Sa(\{T_j, \mathbf{x}_q\})$.

Although more elaborate approaches are also available (Loth and Baker 2013), the conventional way to represent this correlation is to combine inter-period correlation models (Baker and Bradley 2017; Baker and Jayaram 2008) with spatial correlation models at a given period (Jayaram and Baker 2009). This Markovian approximation (Goda and Hong 2008) is represented in Eq. 8.1

$$\rho(\{T_i, \mathbf{x}_p\}, \{T_j, \mathbf{x}_q\}) \approx \rho(T_i, T_j) \times \rho(\mathbf{x}_p, \mathbf{x}_q | \max(T_i, T_j)) \qquad (8.1)$$

This approach is conventionally adopted within portfolio risk analyses. Buildings are assigned to discrete building classes, and each class has a fragility curve developed for it that utilises at least one intensity measure as an input. The risk analysis framework uses Monte Carlo simulation to generate spatially-correlated ground-motion fields at individual co-ordinates, and the motions at these coordinates are input to fragility curves to establish the demands for all buildings in each class.

8.2.2 Effects of Spatial Discretization

Consider again the leftmost panel of Fig. 8.1 in which we have multiple buildings of the same class located within a single cell. In reality, each building occupies a different spatial position and will receive its own value of spectral acceleration. These acceleration values will be correlated spatially over the cell because there will be commonalities associated with source amplitudes, wave propagation paths, and site conditions. The particular amplitude experienced by each building depends

---

[1]Here, we are assuming that ground-motions are described by spectral accelerations. Note that T_i and T_j can be equal, to either represent buildings from the same class, or different classes with the same characteristic response period.

upon the particular realisation of the random field as well as its actual location within
the cell. As spatial correlation models show decreasing correlation with increasing
separation distance (Jayaram and Baker 2009), the further a building is located from
the point within the cell where the ground-motion field is defined, the weaker the
correlation. When looking at the variability in intensity measure amplitudes over the
field, current approaches account for the point-to-point correlations between the grid
points in each cell, but do not also account for the additional variability that arises
over a cell. This additional variability can be computed using Eq. 8.2, which makes
use of an effective correlation, ρ_{eff}, for the cell.

$$\Delta\phi(\mathbf{x}) \approx \phi(\mathbf{x})\sqrt{1 - \rho_{\text{eff}}^2} \tag{8.2}$$

In Eq. 8.2, $\phi(\mathbf{x})$ is the within-event standard deviation of motions at the grid point
for the cell.

To compute the effective correlation, consider the generic geometry shown in
Fig. 8.2. The cell has an area of $\Delta x \Delta y$ and the grid point is indicated by the black
dot. In this schematic we use a rectangular cell and locate the grid point in the
geometric centroid, but there is no requirement to do this generally.

The effective correlation is then computed as the expected value of the correlation
for all possible spatial combinations of locations over the cell, as shown in Eq. 8.3.

$$\rho_{\text{eff}}(T_1, T_2) = \frac{1}{\Delta x^2 \Delta y^2} \iiiint \rho(\{T_1, \mathbf{x}_1\}, \{T_2, \mathbf{x}_2\}) dx_1 dx_2 dy_1 dy_2 \tag{8.3}$$

Note that the default approach in traditional studies is to effectively assume perfect
correlation of $\rho = 1$ for the motions over the cell, while the expression in Eq. 8.3

Fig. 8.2 Geometry of spatial
cells for computation of
within-cell correlation
adjustments

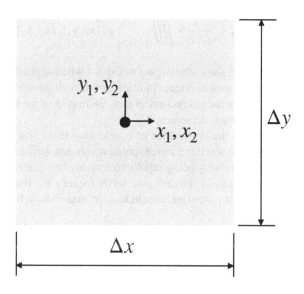

will always be less than unity for any finite cell size. Importantly, for the exponential spatial correlation models that are normally used, the larger the cell size, the smaller the effective correlation.

An important corollary of Eq. 8.3 is that inter-period correlations, that are used to represent correlations among building classes, need to be reduced from their commonly adopted values. Note that when multiple response periods are used as inputs for a fragility function for the same building class, no modification is required as in this case the multiple periods represent multiple attributes of the building at a single location. However, when single spectral ordinates represent different building types, and the exact locations of these buildings are unknown within the cell, we have to reflect the fact that there are many possible combinations of relative locations within the cell that would be associated with different correlation values.

Figure 8.3 demonstrates how the inter-period correlation values of Baker and Jayaram (2008) are modified to account for spatial cell size in a regular square grid of dimension $\Delta x = \Delta y$. One can appreciate that significant reductions in the correlations arise as the nominal cell size increases, i.e., as the spatial resolution decreases.

The next case to consider is the situation where we are interested in the correlations between potentially different intensity measures in different spatial cells. The relevant geometry in this case is shown in Fig. 8.4.

Now the effective correlation is defined by Eq. 8.4, in which the cell sizes are assumed equal for both cells with dimensions $Dx \times Dy$ and the relative positions are defined by Δx and Δy. As before, the sizes of each cell can easily be different, the key concept is that we integrate to ensure that all possible combinations of spatial locations between cells are considered. The $+ = \Delta x, \Delta y$ specification on the integral limits is simply shorthand to denote the relative shift in the x_2 and y_2 co-ordinates relative to x_1 and y_1.

$$\rho_{\text{eff}}(T_1, T_2) = \frac{1}{D_x^2 D_y^2} \iiiint_{+=\Delta x \Delta y} \rho(\{x_1, y_1\}, \{x_2, y_2\}, T_1, T_2) dx_1 dx_2 dy_1 dy_2 \quad (8.4)$$

Figure 8.5 shows the impact of Eq. 8.4 when applied to a regular grid with relative cell offsets equal to integer multiples of the cell dimensions, i.e., $\Delta x = i D_x$ for $i \in \mathbb{Z}$, and similar in the y-direction. Again, the impact of the spatial discretisation increases as the resolution decreases.

Note that the importance of considering these spatially discrete effects is that it allows one to work at a lower spatial resolution whilst still reflecting the appropriate levels of variability being input into fragility functions. In all of the cases considered in this section, as the cell size tends to zero we recover the expressions for the point-to-point cases (and continuous ground-motion fields).

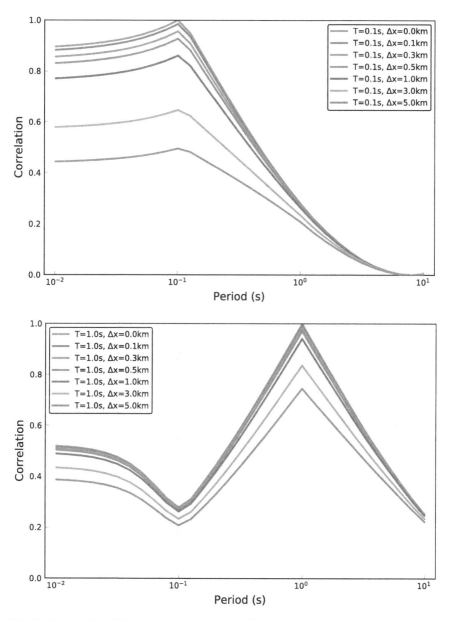

Fig. 8.3 Impact of spatial discretisation size upon the effective inter-period correlations of response spectral ordinates. The upper panel shows conditioning upon a period of 0.1 s, while the lower panel shows conditioning upon a period of 1.0 s

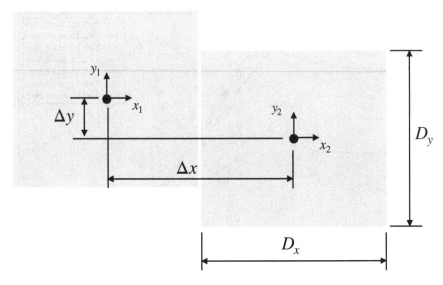

Fig. 8.4 Geometry of spatial cells for computation of between-cell correlations

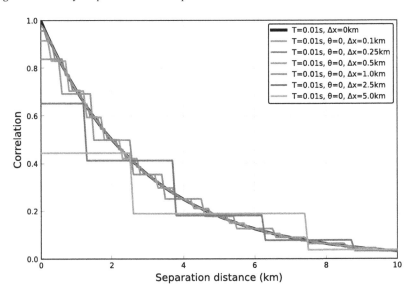

Fig. 8.5 Effective between-cell correlations, accounting for spatial discretisation. $\theta = 0$ indicates that the cells all have the same y co-ordinates and we consider relative positions in the x-direction

8.3 Impact of the Ergodic Assumption upon Correlation Models

As previously mentioned, an ergodic dataset will make use of data from many different spatial locations and ground-motion models derived from this data therefore contain a degree of site-to-site variability that will not exist at a given site location. Correlation models that have been developed in the literature have, for the most part, been computed using a simple partitioning of ground-motion variability into just between-event, δ_B, and within-event, δ_W, components, as shown in Eq. 8.5.

$$\ln im(\mathbf{x}) = \mu(\mathbf{x}; rup) + \delta_B + \delta_W(\mathbf{x}) \tag{8.5}$$

Here, $\mu(\mathbf{x}; rup)$ is the mean logarithmic intensity measure at site x for rupture scenario rup, and we indicate that δ_B and $\delta_W(\mathbf{x})$ are independent, and dependent of position, respectively.

Contrast this with a model in which systematic site effects, $\delta_{S2S}(\mathbf{x})$, are also considered. Now, the event-and-site corrected within-event residuals are represented by $\delta_{W_{es}}(\mathbf{x})$, as shown in Eq. 8.6.

$$\ln im(\mathbf{x}) = \mu(\mathbf{x}; rup) + \delta_B + \delta_{S2S}(\mathbf{x}) + \delta_{W_{es}}(\mathbf{x}) \tag{8.6}$$

Between-event residuals are perfectly correlated (ignoring any parameterisation of nonlinear site effects) for all observations from a given event, so we focus upon the remaining within event correlations.

8.4 Correlations Between Spectral Ordinates at a Point

When deriving correlation models from ergodic datasets, the general expression for the within-event inter-period correlation is given by:

$$\rho(T_i, T_j) = \frac{\rho_{S2S}(T_i, T_j)\phi_{S2S}(T_i)\phi_{S2S}(T_j) + \rho_{SS}(T_i, T_j)\phi_{SS}(T_i)\phi_{SS}(T_j)}{\phi(T_i)\phi(T_j)} \tag{8.7}$$

where $\phi_{S2S}(T)$ is the between site variability at period T, and $\phi_{SS}(T)$ is the single-station variability at period T. Almost all published correlation models are based upon this framework, with only a couple of exceptions (Kotha et al. 2017; Stafford 2017).

As shown in Stafford (2017), the ρ_{S2S} terms are relatively strong and represent different resonance and impedance effects that arise from sites with the same $V_{S,30}$ values. Under a non-ergodic framework in which these systematic site effects are accounted for, the overall correlation changes from $\rho \rightarrow \rho_{SS}$, and to weaker levels of correlation. However, this then requires that the spatial variations of the systematic site terms are evaluated. Currently this is very rare, but at least one regional risk model (Bommer et al. 2017) has attempted this and future applications are sure to move in this direction.

Note that when systematic site effects are accounted for, all of the expressions of the previous section related to spatial discretisation operate on these reduced correlation values. Therefore, we have compounded effects of weaker correlations and discretisation effects. At the same time we have systematic deviations from ergodic median predictions that reflect the systematic site response. Ultimately, what is happening is that we are transferring apparent aleatory variability out of the ergodic ground-motion model and into epistemic uncertainty within a partially non-ergodic model.

8.4.1 Spatial Correlations Between Spectral Ordinates

Turning now to the case where spatial correlations are considered, Eq. 8.8 shows the general expression to define the correlation from correlated random variables $\delta_{S2S}(\mathbf{x})$ and $\delta_{W_{es}}(\mathbf{x})$ at two spatial locations.

$$\rho(\mathbf{x}_i, \mathbf{x}_j) = \frac{\rho_{S2S}(\mathbf{x}_i, \mathbf{x}_j)\phi_{S2S}(\mathbf{x}_i)\phi_{S2S}(\mathbf{x}_j) + \rho_{SS}(\mathbf{x}_i, \mathbf{x}_j)\phi_{SS}(\mathbf{x}_i)\phi_{SS}(\mathbf{x}_j)}{\phi(\mathbf{x}_i)\phi(\mathbf{x}_j)} \quad (8.8)$$

As with models for inter-period correlations at a point, spatial correlation models like Jayaram and Baker (2009) work on within-event residuals according to Eq. 8.5. These models generally use exponential correlation models to represent this spatial variability. In the case of Jayaram and Baker (2009), the authors find that the correlation length depends upon characteristics of the site conditions, namely, whether site conditions are clustered or not. These correlation lengths are shown in Fig. 8.6.

From the framework of Eq. 8.8, it can be appreciated what effect they are really observing. Let the separation distance between two sites be defined as $\Delta = \|x_i - x_j\|$, and assume that exponential correlation models hold for both components of the within-event residuals:

$$\rho_{S2S}(\mathbf{x}_i, \mathbf{x}_j) = \exp\left[-\frac{\Delta}{r_S}\right] \text{ and } \rho_{SS}(\mathbf{x}_i, \mathbf{x}_j) = \exp\left[-\frac{\Delta}{r_W}\right] \quad (8.9)$$

The overall correlation can then be expressed as:

$$\rho(\Delta) = \frac{\phi_{S2S}(\mathbf{x}_i)\phi_{S2S}(\mathbf{x}_j)\exp\left(-\frac{\Delta}{r_S}\right) + \phi_{SS}(\mathbf{x}_i)\phi_{SS}(\mathbf{x}_j)\exp\left(-\frac{\Delta}{r_W}\right)}{\phi(\mathbf{x}_i)\phi(\mathbf{x}_j)} \quad (8.10)$$

Consider limiting cases in which we have full correlation of the systematic site effects $r_S \to \infty$ ($\rho_{S2S} \to 1$), and the case in which we have no correlation at all

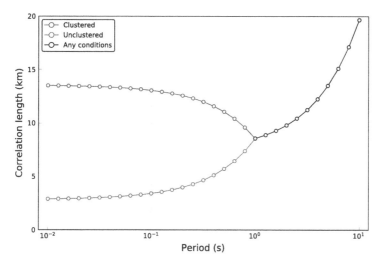

Fig. 8.6 Correlation lengths within the Jayaram Baker (2009) spatial correlation model

among the site effects $r_S \to 0$ ($\rho_{S2S} \to 0$). In the first case, for $r_S \to \infty$ we have:

$$\rho(\Delta; r_S \to \infty) = \frac{\phi_{S2S}(\mathbf{x}_i)\phi_{S2S}(\mathbf{x}_j) + \phi_{SS}(\mathbf{x}_i)\phi_{SS}(\mathbf{x}_j) \exp\left(-\frac{\Delta}{r_w}\right)}{\phi(\mathbf{x}_i)\phi(\mathbf{x}_j)} \tag{8.11}$$

In the second case, for $r_S \to 0$ we have:

$$\rho(\Delta; r_S \to 0) = \frac{\phi_{SS}(\mathbf{x}_i)\phi_{SS}(\mathbf{x}_j) \exp\left(-\frac{\Delta}{r_w}\right)}{\phi(\mathbf{x}_i)\phi(\mathbf{x}_j)} \tag{8.12}$$

The effects of these different conditions, as well as the case where $r_S = r_W$ (equivalent to not decomposing within-event residuals for systematic site effects) are shown in Fig. 8.7. For $r_S \to \infty$ we see that even for very large separation distances we will never tend to zero correlation because we will always have $\rho \approx \phi^2_{S2S}/\phi^2$. Conversely, for $r_S \to 0$ we have a nugget effect as when $\Delta \to 0$ we have $\rho \approx \phi^2_{SS}/\phi^2$. Some studies, such as Stafford et al. (2019), have observed evidence for such nugget effects, but the authors at the time did not fully appreciate the origin of these effects.

As ergodic datasets have different degrees of inherent clustering and hence implicit r_S values, the spatial correlations across site zones can vary significantly (Stafford et al. 2019). When modelling systematic site effects, the above effects need to be taken into account. This point applies both to the derivation of the models in the first instance (taking into account the systematic site terms), as well as during application where the differences in correlations among site zones should be accounted for. Note that for risk analyses working with site zonation models, the spatial correlation

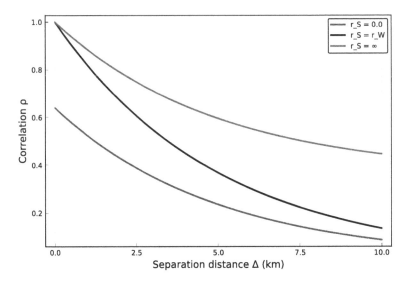

Fig. 8.7 Impact of spatial correlation of systematic site response upon overall spatial correlations

between cells of the systematic site effects should be close to zero (if not actually zero), if the two cells are not in the same zone.

8.5   Non-ergodic Risk Analyses for Seismic Sequences

The final contribution of the present chapter is to discuss issues of non-ergodic ground-motion models relevant for aftershock risk assessments. Studies such as Kuehn et al. (2016); Lee et al. (2020) have shown that systematic source effects from different events can be spatially correlated. However, ergodic datasets rarely have large numbers of events at close spatial locations and so the between-event variability of published models is greater than what should be expected within a single small spatial region. In addition to this, studies (e.g., Kanamori et al. 1993) have discussed the effects that time has upon healing faults and changing frictional characteristics. Therefore, during aftershock sequences, particularly when events are re-rupturing portions of a previously ruptured surface, the frictional characteristics of the rupture surfaces may have less variability than in an ergodic database.

A reasonable working hypothesis is therefore that between-event variability in a small spatial region is lower than the published ergodic values, and that aftershock sequences may have even lower between event variability again. This is important because within a Bayesian updating framework (Stafford 2019) it is possible to actively refine existing ground-motion models as new data becomes available. As a result, aftershock risk analyses can adapt during the sequence to improve risk assessments associated with a given sequence.

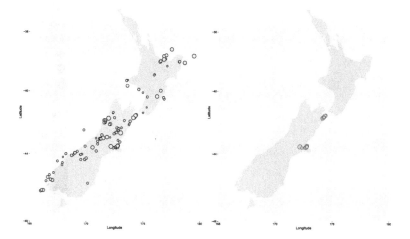

Fig. 8.8 Shallow crustal earthquakes in New Zealand (left), and two key clusters (right) in the Canterbury (red) and Marlborough (green) sequences

To investigate whether we see empirical evidence for this hypothesis, the New Zealand strong ground-motion database is analysed here. On the left of Fig. 8.8, all of the crustal events for which strong-motion records are available are shown. A declustering algorithm (Gardner and Knopoff 1974) is then applied to this data and the two largest clusters of events are extracted. These two clusters correspond the Canterbury and Marlborough sequences and are shown on the right of Fig. 8.8.

A closer view of the spatio-temporal evolution of these earthquake sequences is provided in Fig. 8.9.

For the total database of all crustal events, the NGA-West2 model of Chiou and Youngs (2014) was used to define total residuals that were then partitioned via a mixed effects regression analysis to obtain variance components. The betweenevent residuals for the events in the Canterbury and Marlborough clusters were then extracted and their distribution was compared to the overall between-event variability for the entire database.

Figure 8.10 shows the temporal evolution of the event terms for the Canterbury sequence at two different response periods. The horizontal dashed lines show the total between-event variability for the entire database considered, while the blue lines show loess fits to the data. The shaded region shows the prediction interval for this local fit and it is very clear that this band is significantly narrower than the overall between-event variability.

However, Fig. 8.10 also shows that event terms within the sequence can fluctuate to span a significant portion of the overall ergodic variability.

Similar results can be seen in Fig. 8.11 for the Marlborough sequence. However, in this case we see less temporal fluctuation and a more consistent offset at negative between-event residuals. Of course, just two sequences have been investigated here, but it is important to point out that they have not been identified on the basis of

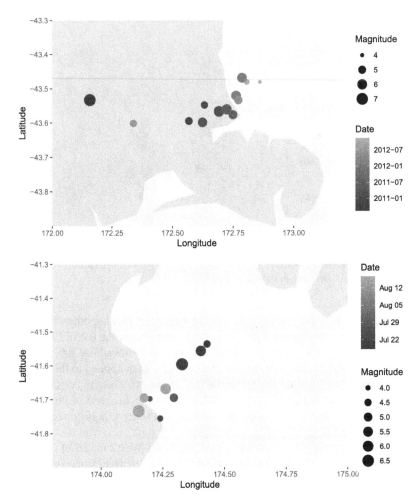

Fig. 8.9 Spatio-temporal evolution of the Canterbury (left) and Marlborough (right) earthquake sequences. Marker size indicates magnitude, while marker shade shows the passage of time from dark being the oldest to light being the most recent

them having any particular characteristics. They are simply the two largest clusters that could be extracted from the available New Zealand strong-motion database. In that sense, the results presented here can be thought of in a similar vein to a blind prediction. That is, a hypothesis was formulated via a thought experiment, and the results obtained are entirely consistent with expectations from that experiment.

In Figs. 8.10 and 8.11, just two periods are shown, but additional summarising results are presented in Fig. 8.12. In Fig. 8.12, the standard deviation of the event terms in the Canterbury and Marlborough sequences are compared to the between-event variability computed from a mixed effects regression analysis using all of the New Zealand crustal data. The standard deviations for the individual sequences are

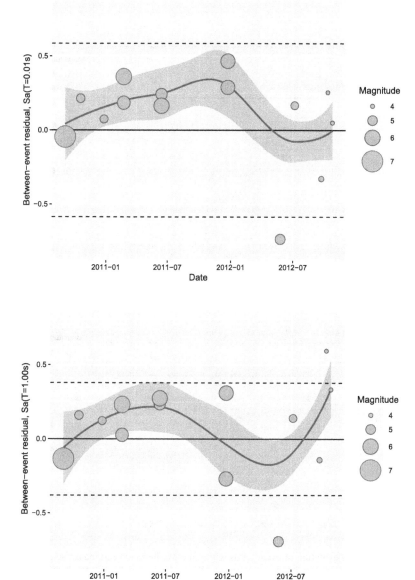

Fig. 8.10 Temporal evolution of event terms within the Canterbury earthquake sequence. The upper panel shows event terms for $T = 0.01$ s, while the lower panel corresponds to $T = 1.0$ s. Markers are sized according to magnitude. Horizontal dashed lines show the ergodic between-event standard deviation for the total database considered. The blue line is a local moving average fit to the event terms and the grey band shows the prediction interval for this curve

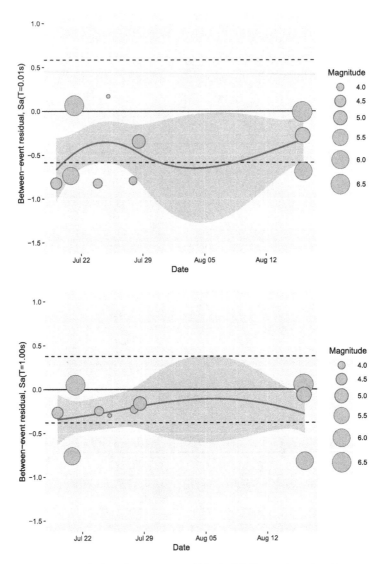

Fig. 8.11 Temporal evolution of event terms within the Marlborough earthquake sequence. The upper panel shows event terms for $T = 0.01$ s, while the lower panel corresponds to $T = 1.0$ s. Markers are sized according to magnitude. Horizontal dashed lines show the ergodic between-event standard deviation for the total database considered. The blue line is a local moving average fit to the event terms and the grey band shows the prediction interval for this curve

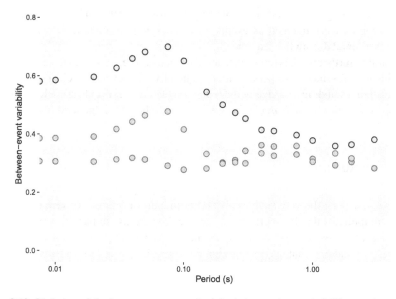

Fig. 8.12 Variation of the between-event standard deviation against period. Blue markers show the variability computed from all New Zealand crustal events from a mixed effects analysis. Red markers show the standard deviation of the event terms in the Canterbury sequence, while green markers correspond to the Marlborough sequence

computed from event terms extracted from the same analysis used to define the overall variability for the entire database.

The results in Fig. 8.12 show a significant reduction at short periods, but it must also be appreciated that it is a significant reduction from a very large level of between event variability for this database. That said, the values for the Canterbury sequence hover around the 0.3 level in natural logarithmic units and this is smaller than typical ergodic values.[2]

It is also important to highlight that these sequences also contain many, many more events than those shown here. Those additional events did not have their strong-motion data processed as part of the New Zealand database analysed here, but in principle a significantly greater amount of data could be available, albeit from small magnitude events, to help constrain the properties of the sequence. Under the assumption that the event terms from the smaller events correlate with those of the larger events, the addition of this weaker motion data could significantly improve one's ability of constrain features of the particular sequence.

This includes overall regional and sequence-specific offsets from ergodic models, as well as systematic site effects. Correlations among these systematic effects, as well as residual correlations can also be updated during the sequence. Within the

---

[2]The total residuals have been obtained from a bias-corrected version of the Chiou and Youngs (2014) model, and this model reports published values of between-event standard deviation that are around 30% greater than what has been found here in the Canterbury sequence.

Bayesian updating framework presented by Stafford (2019), these characteristics can be progressively updated as events occur such that systematic terms become more constrained during the sequence.

Naturally, further work is required to analyse many more sequences to test whether the evidence presented here persists more generally. However, it is clear that some features of these findings, particular the reduction of between-event variability arising from spatially correlated source effects will prove to be a more general finding.

8.6 Conclusions

Regional and portfolio risk analyses have traditionally made use of groundmotion model components that have primarily been derived for use in hazard applications. There are attributes of these components that are not ideally suited for use within risk analyses and this chapter has highlighted some of these issues. In particular, the increasing use of partially non-ergodic approaches within ground-motion modelling has implications for how covariances among intensity measures are represented. The vast majority, if not all, risk analyses currently conducted do not properly account for these effects when attempting to move towards partially non-ergodic approaches. The chapter has shown pathways to address these issues and has also introduced evidence to suggest that withinsequence between-event variability may be over-estimated. This latter point has implications for aftershock risk analyses. However, the potential benefits of working with a reduced variability may be offset by epistemic uncertainty for the earliest events in the sequence.

References

Anderson JG, Brune JN (1999) Probabilistic seismic hazard analysis without the ergodic assumption. Seism Res Letters 70(1):19–28

Baker JW, Bradley BA (2017) Intensity measure correlations observed in the NGAWest2 database, and dependence of correlations on rupture and site parameters. Earthq Spect 33(1):145–156

Baker JW, Jayaram N (2008) Correlation of spectral acceleration values from NGA ground motion models. Earthq Spect 24(1):299–317

Bal IE, Bommer JJ, Stafford PJ, Crowley H, Pinho R (2010) The influence of geographical resolution of urban exposure data in an earthquake loss model for Istanbul. Earthq Spect 26(3):619–634

Bommer JJ, Stafford PJ, Edwards B, Dost B, van Dedem E, Rodriguez Marek A, Kruiver P, van Elk J, Doornhof D, Ntinalexis M (2017) Framework for a ground-motion model for induced seismic hazard and risk analysis in the Groningen gas field. The Netherlands. Earthq Spect 33(2):481–498

Chiou BSJ, Youngs RR (2014) Update of the Chiou and Youngs NGA model for the average horizontal component of peak ground motion and response spectra. Earthq Spect 30(3):1117–1153

Foulser Piggott R, Stafford PJ (2012) A predictive model for Arias intensity at multiple sites and consideration of spatial correlations. Earthq Eng Struct Dyn 41(3):431–451

Gardner JK, Knopoff L (1974) Is the sequence of earthquakes in southern California, with aftershocks removed, Poissonian? Bull Seism Soc Am 64(5):1363–1367

Goda K, Hong HP (2008) Spatial correlation of peak ground motions and response spectra. Bull Seism Soc Am 98(1):354–365

Jayaram N, Baker JW (2009) Correlation model for spatially distributed groundmotion intensities. Earthq Eng Struct Dyn 38(15):1687–1708

Kanamori H, Mori J, Hauksson E, Heaton TH, Hutton LK, Jones LM (1993) Determination of earthquake energy release and ML using Terrascope. Bull Seism Soc Am 83(2):330–346

Kotha SR, Bindi D, Cotton F (2017) Site-corrected magnitude- and regiondependent correlations of horizontal peak spectral amplitudes. Earthq Spect 33(4):1415–1432

Kuehn NM, Abrahamson NA, Baltay A (2016) Estimating spatial correlations between earthquake source, path and site effects for non-ergodic seismic hazard analysis. In: Annual Meeting of the Seismological Society of America. Reno

Landwehr N, Kuehn NM, Scheffer T, Abrahamson N (2016) A Nonergodic groundmotion model for california with spatially varying coefficients. Bull Seism Soc Am 106(6):2574–2583

Lee RL, Bradley BA, Stafford PJ, Graves RW, Rodriguez-Marek A (2020) Hybrid broadband ground motion simulation validation of small magnitude earthquakes in Canterbury. New Zealand. Earthq Spect 36(2):673–699

Loth C, Baker JW (2013) A spatial cross-correlation model of spectral accelerations at multiple periods. Earthq Eng Struct Dyn 42(3):397–417

Rodriguez-Marek A, Rathje EM, Bommer JJ, Scherbaum F, Stafford PJ (2014) Application of single-station sigma and site-response characterization in a probabilistic seismic-hazard analysis for a new nuclear site. Bull Seism Soc Am 104(4):1601–1619

Stafford PJ (2012) Evaluation of structural performance in the immediate aftermath of an earthquake: a case study of the 2011 Christchurch Earthquake. Int J Forensic Eng 1(1):58–77

Stafford P (2014) Crossed and nested mixed-effects approaches for enhanced model development and removal of the ergodic assumption in empirical ground-motion models. Bull Seism Soc Am 104(2):702–719

Stafford PJ (2017) Interfrequency correlations among Fourier Spectral ordinates and implications for stochastic ground-motion simulation. Bull Seism Soc Am 107(6):2774–2791

Stafford PJ (2019) Continuous integration of data into ground-motion models using Bayesian updating. J Seism 23(1):39–57

Stafford PJ, Zurek BD, Ntinalexis M, Bommer JJ (2019) Extensions to the Groningen ground-motion model for seismic risk calculations: component-to-component variability and spatial correlation. Bull Earthq Eng 17(8):4417–4439

Chapter 9
Seismic Fragility Relationships for Structures

L. Di-Sarno and A. S. Elnashai

9.1 Definition and Importance

Structural fragility assessment is a fundamental component of modern performance-based earthquake design and assessment processes. Major advances in fragility functions development and implementation have occurred over the past three decades.

Seismic reliability should be investigated probabilistically via Fragility Functions (FFs) that express the conditional probability of reaching or surpassing a specific damage state given an Intensity Measure (IM) of earthquake shaking. Although damage probability matrices can be used to express structural fragility, a FF is conventionally represented graphically so that an engineer, a stakeholder or a policy maker may be able to visualise the vulnerability of different structural systems. FF also depicts the degree of uncertainty associated with the damage limit state, represented by the shape of the function compared to a vertical line passing through the IM.

FFs constitute an essential step in a consequence-based engineering whereby intervention measures are based on the consequences of reaching or exceeding a certain performance limit state. For example, FF can be utilised prior to an earthquake to devise mitigation and emergency response plans and in the aftermath of an earthquake to prioritise inspection and determine medium- and long-term response and recovery (Elnashai and Di Sarno 2015). Vulnerability functions that correlate

L. Di-Sarno (✉)
Department of Civil Engineering and Industrial Design, School of Engineering, University of Liverpool, Liverpool L69 3BX, UK
e-mail: luigi.di-sarno@liverpool.ac.uk

A. S. Elnashai
Department of Civil and Environmental Engineering, University of Texas at Houston, Houston, USA

© The Author(s) 2021
S. Akkar et al. (eds.), *Advances in Assessment and Modeling of Earthquake Loss*,
Springer Tracts in Civil Engineering,
https://doi.org/10.1007/978-3-030-68813-4_9

the IM with economic losses can be further developed using structural fragility functions and utilised, for example, within insurance schemes at regional or global levels (Pitilakis et al. 2014). Seismic design guidelines could incorporate economic loss models within a life-cycle cost assessment framework that can be used for deciding whether the additional cost due to structural strengthening is a more suitable choice compared to the induced losses by a seismic invent (Calvi et al. 2006). Additionally, the evaluation of risk ensures the uninterrupted operation of a community and is assuming a role of increasing importance due to the increasing complexity and inter-dependence of urban support systems. The main reasons for deriving seismic FFs are summarised in Fig. 9.1.

After providing a brief description of FFs that are used to capture earthquake response data of structural systems, a framework for deriving analytical FFs, which constitute the most widely-used probabilistic method for characterising the probability of failure, is described. Dynamic analysis methods and Engineering Demand Parameters (EDPs) used to describe the response of buildings and bridges are examined by reviewing previous studies. Further insights are gained by reviewing two Case Studies (CSs).

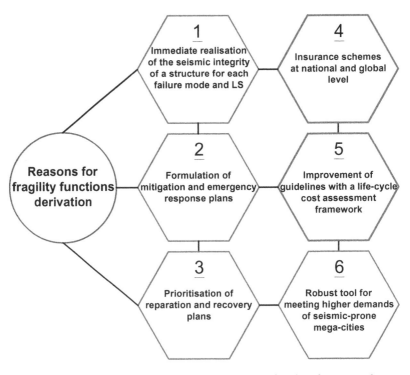

Fig. 9.1 A diagram with the main reasons for deriving fragility functions for structural systems

9.2 Types of Fragility Functions

Different types of FFs exist depending on the way that data are collected (Schultz et al. 2010). First, empirical FFs are formed using observational data that are systematically monitored, controlled and stored. Judgemental data refer to expert opinion and are used as a last resort when observational data are not available. The data may include different modelling parameters; however, the quality strongly depends on the consultant engineer's experience and the bias cannot be as easily controlled. Thus, the empirical method is considered more realistic, whereas the former can account for several structural response factors observed during post-earthquake surveys. Empirical FFs tend to be scarce due to the limited number of data that are primarily at the low seismic intensity range.

Analytical FFs are constructed through mathematical models and can encompass different structural configurations, built environments, geotechnical and seismotectonic characteristics of a seismogenic area. Although analytical FFs can minimise bias referring to material and seismic uncertainty, include all possible failure modes, and yield robust reliability assessment, due consideration should be given during modelling in virtue of software modelling limitations. Since the modelling process can be demanding and onerous, the validity of analytical FFs can be verified with other pertinent FFs. However, this is not always possible given dissimilarities in structural layout, soil properties and seismic input.

For the aforementioned reasons, hybrid fragility investigation can be conducted to compensate modelling difficulties, combine different data sources, and ensure the least possible modelling and seismic uncertainty (Elnashai et al. 2004). Due to the deficiency of observational data for different structural configurations, the empirical or hybrid method is not commonly adopted. However, there are cases where the reliability of observational data is strengthened through analytical studies. These cases primarily pertain to a cluster of structures in seismic-prone areas (e.g. a collection of buildings and bridges in which excessive computational cost is required). A summary of the advantages and shortcomings of each fragility type is presented in Table 9.1.

9.3 Framework for Analytical Fragility Derivation

The analytical approach, which is based on damage distributions derived from the analysis of structural models under incremental seismic intensity, is the most common method of risk assessment. A number of critical steps and assumptions should be carefully followed to analyze the seismic response of a structure, derive the damage distribution, and illustrate the fragility curve. A general framework that clearly encompasses all of the main steps required for evaluating analytical fragility functions is presented in Fig. 9.2.

The type of structure under investigation affects the choice of an analysis software that should take several modelling parameters into account (e.g. material, linear

Table 9.1 Primary advantages and shortcomings of each fragility type

Type	Pros	Cons
Judgemental	Based on expert opinion ample modelling factors for each on-site visit	Inapplicable and insufficient experience of consultant engineer on several structural configurations Quality of data cannot be checked easily
Empirical	More realistic Systematic and controlled data	Specific structural types, geotechnical environment and seismotectonic Scarce data that tend to be narrowed in low seismic intensity range
Analytical	Various and complex structural configurations Reduced bias by accounting modelling and aleatory uncertainties Sensitivity analysis for different layouts and material properties	Not all model types are included in analysis software Bias may still exist due to unrealistic modelling and/or erroneous definition of limit states Onerous and time-consuming
Hybrid	Counterbalance of scarcity and subjectivity of observational data Time efficient for cluster of structures	Functions reliability requires data collection from multiple sources

Table adapted from Schultz et al. (2010)

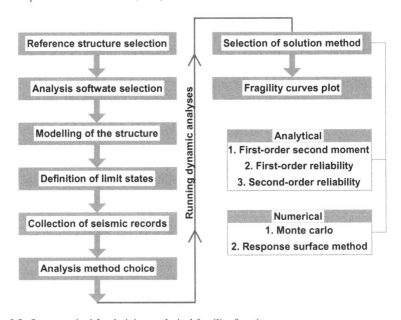

Fig. 9.2 Steps required for deriving analytical fragility functions

and/or nonlinear geometry as well as concentrated or distributed plasticity). Within a fragility analysis framework, it might be time-efficient to adopt simplified yet robust modelling, since the seismic analysis sometimes requires excessive computational capacity depending on the scale and complexity of structural model. In addition, the record-to-record randomness causes higher dispersion of response compared to epistemic/modelling uncertainty, particularly at lower damage states (Kwon and Elnashai 2006; Dolsek 2009; Vamvatsikos and Fragiadakis 2010). Therefore, the definition of limit states and selection of seismic records should be made with due consideration towards estimating robust FFs. Before determining limit states, all possible failure modes should be identified (e.g. local buckling or failure under shear). Subsequently, the performance criteria could regard either the response at the local level (e.g. shear, moment or combined actions) or the global level (e.g. chord rotation, Interstorey Drift Ratio (IDR) or peak floor acceleration as an IM) (ASCE 41–13 2013). Furthermore, a sufficient number of records should be considered to account for geotechnical as well as seismotectonic characteristics of a site. The selection of an analysis method deliberately succeeds the collection of representative seismic records, because depending on the number of available seismic records, the appropriate method can be selected (Shome and Cornell 1999; Jalayer 2003; Bakalis and Vamvatsikos 2018; Di Sarno and Karagiannakis 2020). Selection of an analysis method also depends on the scale of structure, computational capacity and performance target. This step can be considered the final step prior to running dynamic analyses.

The methods to process random variables of seismic response are categorised as analytical and numerical (Schultz et al. 2010; Elnashai and Di Sarno 2015). Numerical solutions commonly assume normal or non-normal distribution of variables with linear limit state equation. Numerical solutions are used when the limit state function cannot be expressed in a closed form, to increase reliability, and to decrease computational time in some cases. Once the limit state function is known, the seismic IM is illustrated as a function of probability of failure for each LS.

9.4 Analytical Fragility Derivation

The modern performance-based engineering framework requires the evaluation of structural reliability in a robust manner. The ease and efficiency by which the data are generated through dynamic analysis of structural models constitute analytical fragility functions as an increasingly attractive method. Although these models can identify bias stemming from modelling and seismic record variability by conducting a sensitivity analysis, they might involve substantial computational effort. Limitations in the modelling capabilities of analysis software may also influence the reliability of this method. The key point of analytical methodologies is the verification of results either with experimental results, which may be limited due to the high cost, or with other analytical studies that use an identical, similar or simplified structural configuration. A challenging task pertains to the evaluation of uncertainties, both epistemic and aleatory, which might be an important contributor in the overall dispersion of performance and a key point that raises issues regarding the validity of current assumptions for all limit states (e.g. the combination of both uncertainties through the square-root-sum-of-squares in performance evaluation). There are computationally demanding and simplified analysis methods that are used to examine the dispersion of performance using different response parameters. Apart from that, the number of records is essential for sufficiently capturing the aleatory uncertainty, which is also highlighted in the following subsections.

9.4.1 Capacity and Demand Uncertainties

Generally speaking, aleatory uncertainty reflects the variability of an outcome (e.g. seismic response that is explicitly recognised by a stochastic model and it is inherently random), while epistemic reflects uncertainty on parameters of the structural model itself (e.g. floor mass, soil nonlinear behaviour, concrete or reinforcing steel strength that are unknown due to lack of knowledge). As previously mentioned, the former type of uncertainty is more considerable than the latter. To account for epistemic uncertainty, modelling parameters of a structure should be assumed as random variables following a certain distribution. In addition to aleatory uncertainty that can be evaluated directly through Incremental Dynamic Analysis (IDA), Dolsek (2009) combined the IDA method with the Latin Hypercube Sampling (LHS) technique in order to define a set of structural models with varying structural parameters. The curves from IDA and modified IDA (modified because of different structural properties assumed for a RC frame) were compared, and it was deduced that modelling parameters do not affect the response in the range far from collapse LS. However, the median collapse capacity was reduced when epistemic uncertainties were considered in the model. The same outcome was also found by Vamvatsikos and Fragiadakis (2010). However, it should be emphasised that the random variables were properly sampled when the number of structural models was greater than the number of random variables, and the greatest influence on the collapse mechanism was observed from random variables with the highest coefficient of variation. Those variables regarded the initial stiffness and ultimate rotation of plastic hinges of columns. In addition, the research focused only on the epistemic variability without comparing the two types of uncertainty.

Furthermore, Porter et al. (2002) conducted a sensitivity analysis on the effect of both modelling and seismic uncertainty in the overall economic performance of a high-rise RC moment frame in California. The impact was measured in terms of a damage factor, which was illustrated to be influenced mostly by the uncertainty in structural capacity and shaking intensity measured in terms of spectral acceleration at the first mode period, as also shown by Kwon and Elnashai (2006). In contrast with Dolsek (2009), uncertainty in the force–deformation relationship was less impactful on the seismic response because of the refined plasticity model that caused a smaller coefficient of variations. Other modelling parameters (e.g. mass and damping) yielded a slight impact on the performance.

The selection of a sufficient amount of records to capture the record-to-record variability is a key component of a seismic reliability analysis. Every study on seismic fragility should be accompanied by the confidence interval for each IM level. For instance, Shome and Cornell (1999) showed that the minimum number of records (typically 3–7) proposed for the professional practice in American and European codes can introduce up to 30% standard error for the one-sigma confidence band of normal distribution. It is generally acknowledged that 10–20 records are adequate for the analysis of low- and mid-rise buildings.

9.4.2   Dynamic Analysis Methods

The fragility analysis methods are categorised into narrow- and wide-range methods depending on the range of IM and displacement values for which they provide demand estimations. The former type of methods can predict the seismic demand e.g. at an IM in the area close to the tolerable probability of a structure, thus they may not accurately predict the variability of records for different performance levels (Jalayer 2003). Single-stripe and cloud analysis are two of these methods. In the first method, a number of records is scaled up to the same intensity, which usually pertains to the exceedance of a predefined limit state. To improve the accuracy of seismic demand prediction, another stripe of records can be formed close to the previous one (usually the initial IM increases by ¼ or ½ of seismic demand dispersion, generally termed β). Furthermore, the cloud analysis is an easily implementable, time-efficient and accurate method. This method provides a cloud rather than a stripe of response values. Thus, the analysis is conducted either with unscaled (as recorded) or scaled records at different IMs. The type, number and intensity of records is decisive for the robustness of the cloud method. The method accuracy lies in estimating the dispersion of seismic demand at each performance level to avoid considering the same slope of regression line at all levels. Since it has been shown that selecting unscaled records may underestimate the seismic demand, record scaling is preferable. According to Jalayer et al. (2017), the records should be scaled so that they cover a wide range of spectral acceleration values in the region of interest, with more than 30% of records exceeding the target limit state and no more than 10% of records pertaining to the same seismic event. While this method requires less computational cost, the number of records should be adequate in order to avoid wide confidence bands from occurring. Using fewer recorded motions is especially helpful in regions where few recorded events exist.

The most common analytical approach is IDA (also called dynamic pushover), where a suite of records is step-wisely amplified, resulting in response curves (or IDA curves) that parameterise the intensity level with the EDP (Vamvatsikos and Cornell 2002). The IDA curves provide a clear picture of seismic response at all performance levels, from yielding until structural instability. This method is the most popular method in fragility assessment, since it is simple in implementation and can give a considerably accurate prediction of structural response. However, the excessive computational effort required to perform hundreds of analyses is considered a deterrent in using this method, apart from being more time efficient. In addition, the scaling of low seismic intensity records at high levels is debatable. To this effect, Baker (2015) suggested the truncated IDA method that requires record scaling up to a maximum IM, independently of whether they have caused collapse or not. With this method, the fragility is estimated by lessening the computational cost and scaling the records up to practical levels.

In contrast with the IDA, the Multiple-Stripes Analysis (MSA) method is performed at specific IMs, each of which has a unique set of ground motions (Jalayer and Cornell 2009). Both MSA and IDA can be characterised as wide-range assessment methods, since they can be conducted for a large range of IMs. A competitive edge of MSA versus IDA is the accuracy due to the compatibility of records with the conditional spectrum at different IMs, since the target properties of records change at each IM. The MSA is also computational demanding given that each stripe should include a considerable number of records which may not be always be possible to find.

The selection of one method over another depends on the single case. To compare the accuracy and time-efficiency between methods, Mackie and Stojadinovic (2005) obtained almost the same EDP—IM relationship as obtained by IDA and cloud method using the same number of records. For example, the IDA method underestimated the median drift ratio by 12%, which may have occurred due to convergence issues caused by significant amplification (e.g. scale factor from 10 to 25). Additionally, a single stripe of records scaled at the IM pertaining to the same seismic hazard further underestimated the median. As a result, the cloud method was proved to yield the best choice or the most conservative response. Evidently, the single stripe can be the best method only when the estimate at a specific IM is required, since it is independent from any error introduced from a mathematic form.

A comprehensive study having as the main goal to compare the results from IDA, cloud and MSA, and to propose a new method that attains the same amount of accuracy with considerably less amount of seismic analyses was proposed by Miano et al. (2018). The concept behind this method, which is called "cloud-to-IDA", lies in accurately obtaining the spectral acceleration value that corresponds to the exceedance of a specific limit state. The regression line is formed by unscaled records, and subsequently, the records are scaled so that the acceleration is close to the acceleration found from the regression line to exceed the limit state of interest. This scaling process is facilitated by forming a box area using the standard deviation of the IM and EDP. Additionally, the largest set of records adopted for this method included 50 initial unscaled records, 19 of which were scaled two times (88 in total) in order to find the acceleration that causes exceedance. The fragility was estimated using 8 times fewer records than MSA that was considered as the "true estimate", since more stripes were located in the area that the exceedance of limit state had found in advance by the classic cloud. The limitation of the proposed method is that it is applied only to one LS. However, the extension of the method to additional limit states is possible

Table 9.2 Primary privileges and shortcomings of different analysis methods in the literature

Analysis method	Pros	Cons
Cloud	Time efficient Use of simple regression Estimation of record-to record variability	High dependence of regression prediction on the suite of records Constant conditional standard deviation of damage given IM
Cloud-to-IDA	Higher accuracy in the true estimate of limit state exceedance Less influence by the suite of records Higher accuracy of standard deviation (the pros of classic cloud are not mentioned to avoid repetition)	Effort for locating a sufficient number of records in the interest area
IDA	Simple in implementation and record selection Thorough understanding of response-IM relation, global system capacity and record-to-record variability Insight into the IM effectiveness	Time consuming Scaling of low magnitude motion may not be accurate Scaling of records up to impractical IMs Bias in the response prediction due to convergence problem in case of highly scaled records
MSA	Wise selection of IMs based upon the dispersion Appropriate for spectrum compatible records for each LS Use of unscaled time-histories Estimation of uncertainty propagation	Not clear picture of collapse capacity Unavailability of records per each IM

by increasing the number of records and the range of spectral acceleration values. In such cases, the total number of records compared to the wide-range method could be smaller, although it still remains to be examined whether the whole process would be less tedious and time effective. Finally, comparing the cloud and IDA method between Mackie and Stojadinovic (2005) and Miano et al. (2018) highlights that the classic cloud method overestimates the capacity. In addition, the choice of the number of records is decisive to avoid any bias from convergence problems that caused twice as much error of media estimation for the former compared to the latter case. The primary advantages and shortcomings of all methods are summarised in Table 9.2.

9.4.3   Solution Methods

A deterministic scenario in which the violation or not of a limit state is signified by probability equal to unity or zero is a simple example of lack of uncertainty. Thus, the fragility becomes a step function with zero or one probability (Fig. 9.3). In this

Fig. 9.3 The step function (deterministic scenario) and S-shaped function with lower and higher probability (after Elnashai and Di-Sarno 2015)

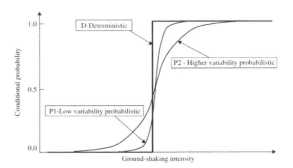

case, the traditional design and assessment process incorporates uncertainty through safety factors that pertain to a specific seismic intensity. Thus, no information is provided for the probability of exceedance at a different intensity level.

In stark contrast, when uncertainty is considered at all levels of intensity, the well-known S-shaped function is formed (Fig. 9.3). The FF of an IM is actually a summation of all structural analysis results conditioned on an IM, and it is expressed as follows (Bakalis and Vamvatsikos 2018):

$$F(IM) = P[EDP > EDP_C | IM] \tag{9.1}$$

where the condition signifies the violation of a certain limit state (EDP_C stands for the capacity of engineering demand parameter). The simplest way of estimating the probability of exceedance of the condition in Eq. 9.1 is the consideration of one EDP_C without uncertainty in its definition. Thus, calculating the ratio between the number of records that violate the condition at each stripe (either from MSA or IDA) over the total number of records, it is possible to estimate the probability at each IM. In this case, it holds that:

$$F(IM) = \frac{\sum_{j=1}^{n} I[EDP^j > EDP_C | IM]}{N_{rec}} \tag{9.2}$$

This process can be considered as a Monte Carlo Simulation (MCS) using n records for each strip, where the index $I(\cdot)$ is an index function taking the value of 1 if the condition is true, and zero otherwise. This is the so-called EDP-basis fragility estimation, since the exceedance of a limit state relies on the EDP_C. The fragility estimation can also be derived based upon the IM_C which is an inherent probabilistic quantity that can be found from hazard maps for a specific site and annual probability of seismic intensity exceedance. After defining the points of IM versus probability of limit state exceedance, it is possible to simply connect the points to form an empirical distribution estimate.

In a different way, an analytical distribution function can be used. The most common analytical distribution function is the Lognormal Distribution Function (aka

CDF), since it has been confirmed as a reasonable assumption [(Ibarra and Krawinkler 2005; Jalayer 2003), among others]. It should be noted that the assumption of any other distribution is another source of uncertainty that can be addressed either with empirical data or by checking the mean annual frequency of exceedance. The CDF is expressed by:

$$P(EDP > EDP_C | IM) = \Phi\left(\frac{\ln EDP(IM)_{50\%} - \ln EDP_{C,50\%}}{\beta_{EDP|IM,tot}}\right) \quad (9.3)$$

where EDP(IM)$_{50\%}$ is the value of the median (50% percentile) at each IM from the IDA curves and EDP(IM)$_{C,50\%}$ is the median. To account for the uncertainty in the capacity (β_C) and damage level definition (β_{DL}), the total dispersion $\beta_{EDP|IM,tot}$ is calculated as follows:

$$\beta_{EDP|IM,tot} = \sqrt{\beta_{EDP|IM} + \beta_C + \beta_{DL}} \quad (9.4)$$

Typical values of the last two dispersions of Eq. 9.4 can be found in HAZUS (2010). The dispersion of seismic demand, $\beta_{EDP|IM}$, is estimated through a lognormal fitting. In case of the IDA method, for instance, the moments can be found by using the natural logarithm of the 16th, 50th and 84th fractiles of EDP. An alternative way to estimate the moments is the maximum likelihood estimation method (Baker 2015) that applies for different types of distribution. If p_j is the probability of observing a collapse, according to binomial distribution, it holds:

$$P(z_j collapses) = \binom{n_j}{z_j} p_j^{z_j} (1 - p_j)^{n_j - z_j} \quad (9.5)$$

where P($\cdot$) is the probability of observing z_j collapses out of n_j records of a stripe. To account for m levels of IM, the product of all probabilities is calculated as follows:

$$total\ likelihood = \prod_{j=1}^{m} \binom{n_j}{z_j} p_j^{z_j} (1 - p_j)^{n_j - z_j} \quad (9.6)$$

The scope of this process is to maximise the total likelihood of Eq. 9.6. This can be done by substituting p_j with a distribution function (e.g. lognormal), and estimate numerically the moments of the function. It is convenient to obtain the derivative of Eq. 9.6 for finding the maximum likelihood.

In the case of a cloud analysis, a logarithmic linear regression is commonly performed to fit the EDP-IM relationship, which is characterised by the following properties:

$$EDP = (a \cdot 1M^b) \cdot \varepsilon$$
$$\mu_\varepsilon = 1\ \&\ \sigma_{\ln a} = \beta_{EDP|IM} \quad (9.7)$$

where *lna* is an intercept and b is the slope in log-space. The lognormal random variable ε has median, μ_ε, equal to unity and its logarithmic standard deviation, $\sigma_{\ln\varepsilon}$, is equal to the standard deviation of natural logarithm of EDP for a given value of IM, $\beta_{EDP|IM}$. As mentioned in Sect. 9.4.2, it is generally recommended that the cloud analysis should be conducted in the region of interest, around the EDP_C, and not in a wide range of IMs. The closed-form solution of FF considering the linear regression of Eq. 9.7 becomes:

$$P(EDP > EDP_C | IM) = \Phi\left(\frac{\ln l M - \ln l M_{C,50\%}}{\beta_{IM|EDP,tot}}\right) \quad (9.8)$$

where the $IM_{C,50\%}$ is the median IM and $\beta_{IM|EDP,tot}$ is given by the value of Eq. 9.4 divided by the value of slope, b. Finally, Bakalis and Vamvatsikos (2018) found that the FFs derived from the closed-form solution of Eq. 9.8, the lognormal fit of Eq. 9.3 and the empirical curve from the MCS (Eq. 9.2) were coincident, which signifies the robustness of the solution methods, either analytical or numerical. It was also deduced that the IM-based method based on IDA analysis is more robust against an EDP-based method, because the dispersion $\beta_{EDP|IM,tot}$ becomes undefined when the first collapses appear.

At this point, it is necessary to discuss two important assumptions that were previously mentioned. First, it should be kept in mind that only one EDP was considered enough to describe the global seismic response of a structure. Even though this consideration is sufficient in most of cases, it might not be adequate when complex structures (e.g. pipe racks or tanks that may exhibit different failure modes) are considered. Second, uncertainty on EDP capacity should also be accounted for beyond the fixed value proposed in HAZUS (2010). For instance, this can be done by combining the IDA method and LHS technique (Dolsek 2009).

Finally, there are additional analytical and numerical solution methods that can be used to derive FFs. This includes the first-order second-moment, first-order (FOSM) reliability or the response surface method. The interested reader can find more information in Iervolino et al. (2004), Schultz et al. (2010), Elnashai and Di Sarno (2015).

9.5 Performance Parameters, Intensity Measures and Applications

The scope of this section is to provide a few examples of the aforementioned analysis and demonstrate the main EDPs that have been adopted for various structural frames as a function of efficient IMs. It should be noted here that a standard deviation constitutes a metric of the efficiency of the regression of tested IMs to describe the seismic response of a structure. Lower $\beta_{EDP|IM}$ values indicate reduced dispersion or a more efficient IM. For instance, Shome and Cornell (1999) addressed the efficiency of several IMs (e.g. PGA, first-mode spectral acceleration (S_a) or averaged S_a over a range of frequencies) as a function of the number of records and scaling method using the cloud method. According to the results, the standard deviation of IDR can be reduced by half when normalizing records at the median S_a. This resulted in the reduction of the number of records by a factor of four given a certain confidence level. Additionally, a number of 10–20 records was proved sufficient to describe the seismic response of a mid-rise building. The study of Miano et al. (2018) is an extension of the previous one, since the authors achieved to reduce the required number of records by scaling them close to the spectral acceleration corresponding to a target LS. This was also confirmed by Jalayer (2003) who investigated the nonlinear response of a RC using IDR and S_a. The one stripe analysis was enough to estimate the seismic demand far from collapse; however, in the near collapse region, two-stripes were necessary to find the true estimates.

Apart from efficient, an IM should be sufficient in that it is conditional independent of seismological characteristics, such as the magnitude (M) and epicentral distance (R). The sufficiency of an IM can be quantified by deriving the p-values of the residuals, $\varepsilon_{EDP|IM}$, based on the regression analysis of EDP with respect to IM. The p-values are derived relative to the M and R. Luco and Cornell (2001) examined the efficiency and sufficiency of S_a, spectral displacement (S_d), and a modified spectral acceleration considering the second-mode period contribution and inelasticity (S^I_{a2}). The IMs were examined with respect to IDR for moderate-to-long period structures. The main outcome of the study highlighted that the S^I_{a2} was the most appropriate IM to describe the seismic response of a 3-, 9- and 20-storey building. As such, the seismic records characteristics can be ignored when employing this IM.

As mentioned in Sect. 9.4.1, the epistemic uncertainty could impact the seismic demand close to collapse LS. Dolsek (2009) analysed a 4-storey RC structure employing the LHS technique to create samples with different mechanical characteristics. To avoid problems in the definition of S_a due to the different period of the structure for each sample, the PGA was adopted and correlated with the maximum IDR. The dispersion of PGA and drift demand due to randomness was estimated at 0.68 and 0.46, whereas it reached the value of 0.79 and 0.56, respectively, when both types of uncertainty were considered. Instead of using PGA as IM, Vamvatsikos and Fragiadakis (2010) considered the S_a, since only the strength and not the mass or stiffness varied, thus the period remained the same. However, even under mass and stiffness uncertainties, it has been shown that the S_a can still serve as a reliable reference IM (Vamvatsikos and Cornell 2005).

The fragility analysis of bridges has also been the subject of many studies during the last decade. There has been a lack of agreement regarding the most suitable IM for bridges (e.g. spectral measures versus ground ones). For example, Mackie and Stojadinovic (2003) addressed the probabilistic seismic demand of bridges with 23 different IMs. It was highlighted that structure-dependent IMs (e.g. S_a and S_d at the fundamental period of the bridge) reduced the seismic demand uncertainty. Ground measures (e.g. peak ground velocity or duration-dependent ones such as root mean square acceleration) were not useful. In addition, local, intermediate and global EDPs were examined, such as maximum material stresses (σ), column moment (M) and IDR. Conversely, the research of Padgett et al. (2008) on a portfolio of multi-span simply supported steel girder bridges demonstrated that PGA can be the best contender out of 10 other typical IMs as a function of bearing deformation (br), ductility demand (μ) and abutment deformation (abut). Apart from the efficiency and sufficiency, the authors examined another factor, namely proficiency, which combines both efficiency and practicality. Practicality is defined as the slope b of IM-EDP relation. In this way, different factors that affect the decision-making can be combined for a proper selection of an IM. It was pointed out that the PGA served as the most proficient followed by $S_{a\text{-gm}}$ (geometric mean of two orthogonal principal periods). The cumulative absolute velocity (CAV) and PGA were proved to be the most sufficient measures. Finally, the study confirmed that differences between the proposed IMs in the literature for bridges cannot be attributed to the nature of ground motions, synthetic or recorded, but rather to specific characteristics of individual and portfolios of bridge classes. Other remarkable studies regard the impact of near- and far-field conditions on bridges (De Risi et al. 2017), as well as the consideration of SSI effects (Kwon and Elnashai 2010) that need to be examined more as an epistemic source of uncertainty.

Two critical developments of fragility analysis pertain to the consideration of residual capacity of structures subjected to aftershock events, and the ageing effects of structures (e.g. due to corrosion). First, the damaging effects of aftershock events are overlooked by the design codes. Additionally, fragility analysis of structures usually addresses only mainshock earthquakes, although some structures can be prone to sequence of seismic events. Jeon et al. (2015) confirmed this statement by showing that the PGV of an aftershock required to cause severe damage to a RC frame was 30% lower compared to the one, the frame was undamaged. The study used the IDA approach for simulating damaged ground motions and the cloud approach for computing aftershock FFs. Although PGV was found the most proficient IM, CAV was the most practical. Apart from deterioration related to mainshock-aftershock events, structural degradation may occur due to corrosion that affects concrete cover and steel reinforcement strength. Panchireddi and Ghosh (2019) introduced a novel study on aged RC bridges subjected to mainshock-aftershock sequences. Results showed that corrosion has a significant impact on the seismic vulnerability of RC bridges, which becomes even more critical when the bridge is subjected to both ground motion sequences and harsh corrosion conditions. The most important considerations of the aforementioned studies on fragility are summarised in Table 9.3.

To illustrate the generation of analytical FFs and further investigate the critical subject of the reliability assessment of structures subjected to aftershocks and experienced corrosion, two analytical CSs are addressed in the ensuing sections accounting for IDA and different IMs.

Table 9.3 Primary considerations of several fragility analyses on building structures and bridges

Reference	Structural type	Fragility method	Solution method	Uncertainty	EDP	Efficient IM
Shome and Cornell (1999)	5- and 20-storey steel moment resisting frame	Cloud with confidence levels	Logarithmic linear regression	Aleatory	Global or storey drift, hysteretic energy	Sa or averaged Sa
Luco and Cornell (2001)	3-, 9- and 20-storey steel moment resisting	Cloud	Logarithmic linear regression	Aleatory	IDR	SIa2
Jalayer (2003)	7-storey RC frame	Analytical (IDA, MSA, cloud)	Numerical and analytical (regression)	Aleatory	IDR	Sa
Dolsek (2009)	4-storey RC frame	IDA and modified IDA	Analytical and numerical (LHS)	Epistemic and Aleatory	IDR	PGA
Vamvatsikos and Fragiadakis (2010)	9-storey steel moment resisting frame	IDA and modified IDA	Monte Carlo (LHS, point estimate and	Epistemic and Aleatory	IDR	Sa
Mackie and Stojadinovic (2003)	Single/multiple span RC bridge	Cloud	Linear regression	Aleatory	σ, M and drift ratio	Sa or Sd
Padgett et al. (2008)	Multi-span steel girder bridges	Cloud	Linear regression	Aleatory	μ, abut, br	PGA (Sa,gm, CAV)
Miano et al. (2018)	7-storey RC building	Cloud-to-IDA Comparison of different methods	Logarithmic linear regression	Aleatory	Maximum chord rotation	Sa
Jeon et al. (2015)	4-,8- and 12-storey RC frames	cloud (aftershock) IDA (mainshock)	Analytical and numerical (LHS)	Epistemic and Aleatory	IDR	PGV
Panchireddi and Ghosh (2019)	Two-span RC-bridge with ageing	Cloud	Logarithmic linear regression and MCS	Epistemic and Aleatory	Damage index	PGA

9.6 Aftershock Fragility Analysis of a Steel Frame (CS#1)

9.6.1 Description

The present case study (CS) demonstrates a simple framework of implementing aftershock fragility analysis on existing steel frames. The case study is a three-storey existing steel moment-resisting frame located in Central Italy, which has a trapezoidal floor plan and a storey height of approximately 3.6 m for the three storeys. Figure 9.4 shows the plan layout of the steel building. The external and internal beams are HEA160 and HEA300, while the columns are HEA200. All beams were found to be connected to columns through full penetration welds. Lastly, the masonry infill walls consist of two layers of perforated bricks of thickness 60 mm.

The numerical model of the bare and infilled frames were implemented in OpenSees (Mazzoni et al. 2006). Beams and columns were modelled as force-based elements with fibre sections, whose property was represented by the Giuffré-Menegotto-Pinto constitutive law (Menegotto and Pinto 1973; Filippou et al. 1983). Due to a lack of onsite material tests, the actual yield strength of steel considered for the numerical model was 215 MPa, assuming a standard deviation of 15 MPa and a confidence factor of 1.2, according to the knowledge levels defined in EC8-3 (EN 1998–3 2004). The elastic modulus of the steel was 210GPa with a strain hardening ratio of 0.02. Aside from the beams and columns, the column panel zones were also accounted for in both models of the bare frame and the infilled frame. The modelling of column panel zone was developed by Gupta and Krawinkler (1999), which physically modelled the rectangular shape of the column panel zone through small rigid elements and utilised a rotational spring to control the shear deformation of the column panel zone. The modelling of the column panel zones was employed in both the bare frame and the infilled frame. Masonry infilled walls were modelled using the single-strut model due to its simplicity and acceptable accuracy. The infill struts had the same thickness as the real infilled walls, and their width was determined based on the properties of the infill walls and the confining frame (Noh et al. 2017). The

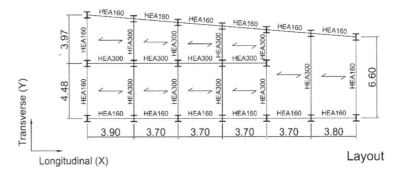

Fig. 9.4 Layout of the case study steel building

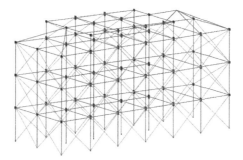

Fig. 9.5 3D model of the steel building in OpenSees (slab elements omitted for clarity)

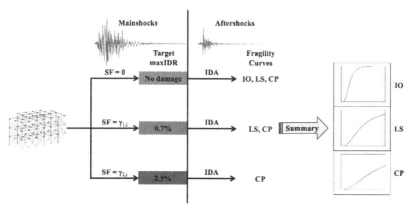

Note: $\gamma_{1,i}$ and $\gamma_{2,i}$ are scaling factors for the ith mainshock
aiming at the target maximum IDR (maxIDR) of structures.

Fig. 9.6 Procedure for assessing steel frames with different levels of pre-existing damage

backbone curve of the masonry infill strut was represented by the multi-linear curve developed by Liberatore and Decanini (2011). Finally, the floor slab on each storey was simplified as two rigid struts placed diagonally in each column grid. Figure 9.5 shows a schematic view of the 3D-model of the infilled frame in OpenSees.

9.6.2 *Methodology*

First, a set of 20 bi-directional records of earthquake sequences were selected from worldwide ground motion databases, including PEER (2013), Luzi et al. (2019), K-NET (2019), to be employed in the finite element model. Each earthquake sequence comprises two events (i.e. the mainshocks and the aftershocks in the order of their time of occurrence in reality). Table 9.4 summarises the PGA of the selected earth-

Table 9.4 Selected mainshock-aftershock earthquake records for fragility analysis

Event	PGA (g)		Event	PGA (g)	
	Mainshock	Aftershock		Mainshock	Aftershock
Armenia Spitak	0.269	0.089	Japan Fukushima	0.577	0.192
Chile Valparaiso	0.787	0.265	Japan Niigata	0.683	0.764
China Northwest	0.360	0.190	New Zealand Christchurch	0.287	0.233
Greece Kalamata	0.285	0.309	New Zealand Edgecumbe	0.599	0.132
India Chamoli	0.418	0.078	New Zealand Weber	0.242	0.323
Iran Varzaghan	0.573	0.747	Taiwan Chi-Chi	1.316	0.517
Italy Emilia	0.372	0.369	Turkey Duzce	0.423	0.265
Italy Friuli	0.457	0.143	USA Chalfant	0.600	0.360
Italy Irpinia	0.433	0.104	USA Mammoth	0.610	0.245
Italy Nocera Umbra	0.740	0.609	USA Whittier	0.484	0.280

quake records. It is noticed that only 4 out of 20 records have a greater after-shock PGA than the mainshock PGA. The lowest and highest ratio of aftershock PGA to mainshock PGA is 0.19 and 1.33, respectively.

Furthermore, the damage levels adopted in this CS are recommended in the American code ASCE 41–06 (2005) for existing steel MRFs. The IDR limits are 0.7, 2.5 and 5% for immediate occupancy (IO), life safety (LS) and collapse prevention (CP), respectively. The IO level indicates that slight damage occurs on structures, such as minor cracks on infilled walls, but the effects on vertical load resisting systems are negligible. The LS level means moderate damage on structures, with large cracks on infills, significant yielding in steel components and permanent residual drifts. However, structures still have adequate residual strength to sustain the gravity loads so that partial collapse is prevented. Finally, the CP level suggests that partial or total collapse occurs on structures, with large permanent residual drifts and very limited vertical load carrying capacity.

The analysis framework described hereafter is able to assess the seismic vulnerability of structures with different levels of damage caused by mainshocks (i.e. the pre-existing damage on structures before aftershocks). The maximum inter-storey drift ratio was used as the EDP, while PGA, $S_a(T_1)$ and CAV of aftershocks were used as the IM. In this CS, three pre-existing damage levels caused by mainshocks were considered, namely 'no damage', 0.7% and 2.5% IDR damage level. 'No damage' means that the steel frame was subjected to aftershocks only, 0.7% IDR represents the worst case of 'slight damage' caused by light mainshocks according to the limit states, and 2.5% IDR represents the worst case of 'moderate damage' that is caused by moderate mainshocks. The analysis procedure is summarised in the following steps (see also Fig. 9.6):

- Scale each mainshock individually based on the results in the first part such that the infilled steel MRF reaches the target maximum IDR after the mainshocks;
- Perform IDA on the damaged infilled steel MRF based on the increasingly scaled aftershocks up to the CP limit state;
- Derive aftershock fragility curves for the infilled steel MRF with each of the assumed pre-existing damage caused by mainshocks;
- Examine the seismic vulnerability of the damage steel MRF by comparing the fragility curves using the case of no damage as a reference;
- Use 16th and 50th percentile values to quantify the change of the steel MRF's seismic vulnerability due to pre-existing damage.

9.6.3 Results and Discussion

Figure 9.7 shows the aftershock fragility curves with respect to the LS limit state, and Fig. 9.8 shows the comparisons between the obtained fragility curves, where the case of no damage was used as a reference. It is evident that the 0.7% IDR damage exhibited negligible impact on the seismic vulnerability of the CS steel MRF, suggesting that this pre-existing damage level is too slight to influence the capacity of the steel frame to resist aftershocks. The comparison of 16th and 50th percentile

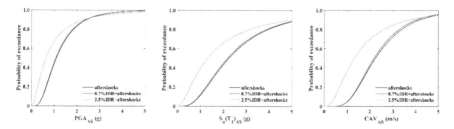

Fig. 9.7 Fragility curves of the steel frame with respect to the LS limit state

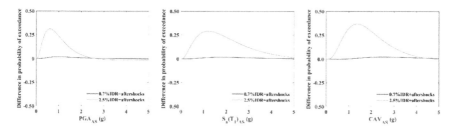

Fig. 9.8 Changes in the probability of exceedance with respect to no pre-existing damage

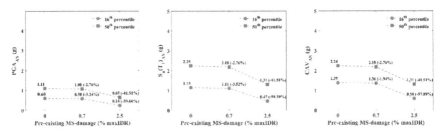

Fig. 9.9 Comparison of the 16th and 50th percentile values for the case of LS limit state

values of IMs in Fig. 9.9 also indicates the slight impact of 0.7% IDR pre-existing damage. For example, when there is no damage that resulted from mainshocks, the steel MRF is believed to reach the LS damage level at aftershock $S_a(T_1)$ of 1.15 g, while in the case of 0.7% IDR pre-existing damage, the structure reaches the same limit state at aftershock $S_a(T_1)$ of 1.11 g, which is 3.5% less than the case of no damage. Similar observations are also found by using other IMs, i.e., PGA and CAV in this CS. Conversely, the 2.5% IDR pre-existing damage has more considerable impact on the seismic vulnerability of the steel frame. Since 2.5% IDR is also the onset of LS damage level, this requires the steel frame to experience a smaller IDR during the aftershock than 2.5%. In this case, the 16th percentile value is an ideal representative of the breakpoint beyond which the aftershocks can cause a larger maximum IDR of the steel frame than the mainshocks.

Figures 9.10 and 9.11 present the results of fragility analysis with respect to the CP limit state. The findings are generally similar to the previous case for the LS limit state. Firstly, the 0.7% IDR pre-existing damage has very limited impact on the seismic vulnerability of the steel MRF. The reduction of the 16th and 50th percentile values of aftershock PGA, $S_a(T_1)$ and CAV in Fig. 9.9 are all less than 2%. It is therefore anticipated that the steel frame with slight pre-existing damage is able to exhibit full capability of resisting aftershocks. When the pre-existing MS-damage was raised to 2.5% IDR, the effects of pre-existing damage becomes more significant. The large changes in the 16th and 50th percentile values demonstrate that the steel frame with moderate damage may have already lost the majority of its capacity to

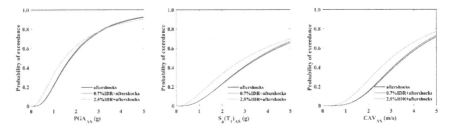

Fig. 9.10 Fragility curves of the steel frame with respect to the CP limit state

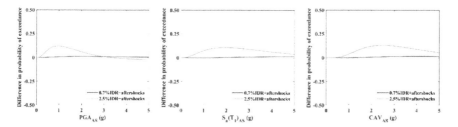

Fig. 9.11 Changes in the probability of exceedance with respect to no pre-existing damage

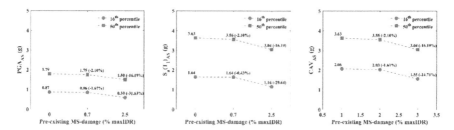

Fig. 9.12 Comparisons of the 16th and 50th percentile values for the case of CP limit state

sustain aftershocks; therefore, there is likely to be a large increase in the structure's seismic vulnerability (Fig. 9.12).

Based on the above assessment, it can be concluded that when the steel MRF is slightly damaged by a mainshock (e.g., the structure is identified between no damage and IO limit state), the steel frame is able to maintain its full capacity to resist aftershocks. As a result, it is not necessary in this case to take the effects of the pre-existing damage into consideration when performing code-based assessment of the steel frame. However, when the steel MRF is moderately damaged by a mainshock (e.g., the structure is identified between IO and LS limit state), the steel frame is likely to lose most of its capacity, which makes the structure significantly more vulnerable to aftershocks. In this case, the effects of the pre-existing damage must be appropriately accounted for during the implementation of code-based assessment procedure, such as a reduction in the criteria for determining capacity or an increase in the seismic action for superior limit states. The amount of such reduction or increment may be effectively determined based on the 16th percentile value of IM to be on the safe side or based on the 50th percentile value to be less conservative.

9.7   Seismic Fragility of a RC Building with Corrosion (CS#2)

There are two main aspects involved with existing RC buildings, namely poor seismic details (e.g. large stirrups spacing or small concrete cover thickness) (Pinto and Franchin 2010; Di Sarno et al. 2017), and deterioration due to exposure to aggressive environmental conditions (corrosion) that alter the most relevant mechanical properties and cause cover spalling, loss of bond between concrete and steel bars, as well as concrete and steel strength reduction, among others (Wang and Liu 2008; Liberatore and Decanini 2011).

Presently, technical codes focus strictly on the design level and even when they deal with existing structures, they limit the checking for strength requirements at the local level without considering and invoking the interaction between elements that are responsible for the structural behaviour and even the structural failure. Although corrosion remains an unpredictable phenomenon, many attempts have been made for incorporating such uncertainties in complicated modelling to allow researchers to account for the life-time deterioration of RC structures. Such an attempt is examined in the following CS by deriving FFs accounting for different corrosion rates.

9.7.1   Description

Non-linear finite element model (FEM) of an existing four-storey RC building was implemented in an advanced software (SeismoSoft 2019) for seismic simulations (Fig. 9.13). The model consists of $350 \times 350\,\text{mm}^2$ and $300 \times 300\,\text{mm}^2$ columns at the ground floor and the remaining floors, respectively, reinforced with 6 smooth $\Phi 16\text{mm}$ longitudinal rebars and $\Phi 6\text{mm}$ transverse stirrups with 150 mm spacing, while beams had variable cross-sections, reinforced mainly with $\Phi 10\text{mm}$ and $\Phi 14\text{mm}$ longitudinal rebars, and $\Phi 6\text{mm}$ transverse bars with 200 mm spacing. The concrete compressive strength was 16.7 MPa, as was typical for buildings designed in the 60 s, while the yielding stress of steel reinforcement was 440 MPa. To guarantee an

Fig. 9.13 Finite element Model in Seismostruct

in-plane stiffness and reduce the number of degrees of freedom, and thus the computational demand, the slabs were modelled through rigid diaphragms, to exhibit neither membrane deformation nor report the associated forces. All the end-joints were rigid connections. An accurate estimation of the gravity loading analysis was conducted and applied to the FE model (see Di Sarno and Pugliese 2020 for further details). Corrosion was applied to the edged beams and columns to simulate a real exposure since the internal components are commonly protected by in-fills.

A new approach was shown to be efficient and reliable for the evaluation of the residual capacity of RC components exposed to corrosion. The methods consist of splitting the RC cross-section in three different layers accounting for the concrete cover (CC), the ineffective core concrete (UCC), and the effective core concrete (ECC). The ineffective core concrete is taken as twice the average diameter of the longitudinal steel rebars and affected by corrosion. Conversely, the core concrete is considered pristine and without any effects of corrosion. The last statement has a physical meaning, as experimental results on RC columns exposed to different levels of corrosion demonstrated that the core concrete is not subjected to the corrosion effects (Andisheh et al. 2019). The above numerical method has the advantage to include the effects of corrosion both full-sided and no-full-sided attack. The last observation comes handful when assessing RC buildings, whereas infills protect some edges of beams and columns, while bridge piers are more likely to experience a full-sided corrosion penetration. As a result, the concrete compression strength deterioration can be computed as follows:

$$f_c^* = \frac{\beta f_c A_{CC} + \beta f_{cc} A_{UCC} + f_{cc} A_{ECC}}{A_{CC} + A_{UCC} + A_{ECC}} \quad (9.9)$$

where f_c is the un-corroded concrete compressive strength and A is the area of each concrete layer. β is defined according to Coronelli and Gambarova (2004) using the modified field compression theory of Vecchio and Collins (1986) as follows:

$$\beta = \frac{f_c^*}{f_c} = \frac{1}{1 + K \frac{2\pi X n_{bars}}{b \varepsilon_{c2}}} \quad (9.10)$$

where f_c^* represents the corroded compressive strength, K a constant equal to 0.1 for medium rebar, X the corrosion penetration, b the width of the cross-section, ε_{c2} strain at the peak and n_{bars} the number of steel reinforcement in the area affected by corrosion.

The degradation effects of corrosion on steel reinforcement are commonly considered by modifying the main parameters of the constitutive models such as yielding and ultimate stress, and the ultimate strain. Many experimental campaigns have been conducted on the impact of corrosion on such mechanical properties and as a result, have demonstrated that yielding and ultimate stress can be easily defined by a linear relationship, while an exponential interpolation is more likely to fit the reduction of the ultimate strain. Moreover, it should be stressed that corrosion can be categorised as uniform, most likely as carbonation due to concrete, and pitting, most likely due to chloride ingress. Both corrosion phenomena may have different impacts on the residual capacity of RC components. Pitting corrosion has larger and more unpredictable effects on the steel diameter and its mechanical properties, while uniform corrosion can be modelled efficiently by uniformly modifying the parameters along the rebar. Results from the literature show that regression analyses produced more or less the same relationships, as follows:

$$f_y^* = \left(1 - \beta_{sy} C R[\%]\right) f_y$$
$$f_u^* = \left(1 - \beta_{su} C R[\%]\right) f_u$$
$$\varepsilon_{su}^* = \varepsilon_{su} e^{-\beta_{\varepsilon u} C R[\%]} \qquad (9.11)$$

(parameters with symbol * represent the corroded variables; β_{sy}, β_{su} and $\beta_{\varepsilon u}$ are the regression parameters). Some results for the regression parameters can be found in Wang and Liu (2008), Imperatore et al. (2017).

9.7.2   Methodology

As stated in Sect. 9.4.1, to obtain an adequate and accurate average inelastic response of a low-rise building, 10–20 seismic records should be considered. Thus, a set of 20 natural ground motions were collected from international databases using a REXEL tool (Iervolino et al. 2010). These ground motions show different features in terms of duration, PGAs, fault rupture and frequency contents (Table 9.5). Since the ground motions were selected and employed in the model considering the two components, the response parameters are then computed using the square root of the sum of squares

Table 9.5 Selected mainshock-aftershock earthquake records for fragility analysis

W	M	PGA_x	PGA_y	D_x	D_y	IA_x	IA_y	P_x	P_y
333	6.6	2.26	3.04	15.4	13.8	61.5	8.8	0.52	0.26
1726	6.3	2.16	2.64	13.0	13.2	86.2	96.9	0.66	0.52
439	6.7	1.79	1.80	8.8	10.5	30.2	25.1	0.30	0.36
592	6.0	1.95	2.18	9.7	11.6	47.2	41.8	0.16	0.08
1254	7.6	1.76	1.56	32.2	34.4	90.2	63.8	0.54	0.38
1257	7.6	2.90	2.39	32.1	33.5	146.4	138.0	0.26	0.52
591	5.7	3.30	2.56	5.38	5.4	68.8	53.3	0.18	0.34
581	5.4	1.72	1.96	8.92	9.02	18.81	19.34	0.46	0.40
4343	7.6	1.08	1.12	39.45	38.99	23.98	31.92	0.64	0.44
602	6.0	1.14	1.07	11.73	11.45	8.23	9.58	0.14	0.14

(SRSS) (RS—Response Parameter):

$$RS_{tot} = \sqrt{RS_x^2 + RS_y^2} \tag{9.12}$$

Robust fragility analysis requires an accurate selection of performance levels that account for local and global response for RC structures. Such performance levels should lead to reliable evaluation of force demands on potential brittle failure, quantification of consequences of strength deterioration on single components, estimation of the inter-storey drift to account for strength and stiffness discontinuities. Technical codes (i.e. EN1998-1 (2004); NTC 2018) usually state that existing RC structures should comply with deformation capacity through chord rotation and cyclic shear resistance; however, the last parameters do not take into account the deterioration of materials, components and, as a result, the global structure. In addition to the above performance levels, some other response parameters could be included, such as the strain of the cover, $\varepsilon_{CU,COVER}$, and core concrete crushing, $\varepsilon_{CU,CONFINED}$, the interaction bending moment-axial load domain, ($N_{Y,COLs}$ and $M_{Y,COLs}$), defining specific limitation on the materials, flexural capacity, $M_{U,BMs}$, of RC components through the bending moments. Table 9.6 summarises the local performance levels.

The values of strains for the structural materials were computed according to the studies of Biskinis and Fardis (2009), Razvi and Saatcioglu (1994) for the unconfined and confined concrete. The latter parameters were then used as limits for the calculation of the interaction domain of each RC component. The global response parameters were taken from non-linear static analyses performed considering the different levels of corrosion and picking the ultimate drift from the capacity curves (Table 9.7). Such performance values represent the capacity of the RC building.

While performing the non-linear analyses, the first element reaching the limit conditions is taken by means of the drift. Among those parameters the minimum is then chosen as demand and checked against the corresponding global capacity parameter according to the limit state. The local EDP over the decision global variable is herein taken as critical demand-to-capacity ratio and defined as (De Risi et al. 2017):

$$Y_{LS} = \max_{i=1}^{N_{Mech}} \min_j^{N_{comp}} \frac{D_{ji}}{C_{ji}(LS)} \tag{9.13}$$

where N_{mech} is the number of considered potential failure mechanisms and N_{comp} the number of components taking part in the ith mechanism. D_{ji} is the demand evaluated for the jth component of the ith mechanism and C_{ji} (LS) is the limit state capacity for the jth component of the ith mechanism.

In this study, the fragility assessment is based on three different IMs, namely PGA, $S_a(T_1)$, and the modified acceleration spectral intensity (MASI) that has been recently introduced, and it is defined as follows:

Table 9.6 Performance criteria (DL–limited damage; SD–severe damage; NC–Near Collapse)

Local parameters		
DL	SD	NC
ε_C, θ_C, ε_{SY}, $M_{Y,BMs}$ $\left(N_{Y,COLs},\ M_{Y,COLs}\right)$, V_Y	$\varepsilon_{CU,COVER}$, $\dfrac{3}{4}\theta_U$, $\dfrac{3}{8}\varepsilon_{SU}$, $\left(\begin{array}{c}N_{Y,COLs}\left(\varepsilon_{CU,COVER},\ \frac{3}{8}\varepsilon_{SU}\right),\\ M_{Y,COLs}\left(\varepsilon_{CU,COVER},\ \frac{3}{8}\varepsilon_{SU}\right)\end{array}\right)$, V_{SD} $M_{U,BMs}\left(\varepsilon_{CU,COVER},\ \dfrac{3}{8}\varepsilon_{SU}\right)$	$\varepsilon_{CU,CONFINED}$, θ_U, ε_{SU}, $M_{U,BMs}\left(\varepsilon_{CU,CONFINED},\ \varepsilon_{SU}\right)$ $\left(\begin{array}{c}N_{Y,COLs}\left(\varepsilon_{CU,CONFINED},\ \varepsilon_{SU}\right),\\ M_{Y,COLs}\left(\varepsilon_{CU,CONFINED},\ \varepsilon_{SU}\right)\end{array}\right)$,

Corrosion rate [%]	NC [%]	SD [%]	LD [%]
0	2.30	1.87	1.06
10	1.69	1.44	0.93
20	1.21	1.17	0.90

Table 9.7 Limit States expressed as inter-storey drift ratio, IDR [%]

NC = Near Collapse; SD = Severe Damage; LD = Limited Damage

$$M.A.S.I. = \int_{T_1}^{T_{elongation}} S_a(T)dT \qquad (9.14)$$

The selection of an IM is challenging, as highlighted in Sect. 9.5, let alone when degradation phenomena due to corrosion are considered in this CS. Thus, there is a higher need to investigate different criteria (e.g. efficiency, proficiency and practicality).

The IDA method is adopted for deriving FFs for the RC building. The scaling of the records until collapse was achieved via the hunt-fill algorithm (Vamvatsikos and Cornell 2004), which defines a first elastic start at 0.005 g, an initial step of 0.1 g and a step increment of 0.05 g. After running each record, the $S_a(T_1)$ is estimated based upon the scaling, and the IDA curves are formed using a spline interpolation. Finally, the whole procedure for deriving FFs for the RC building is described in Fig. 9.14.

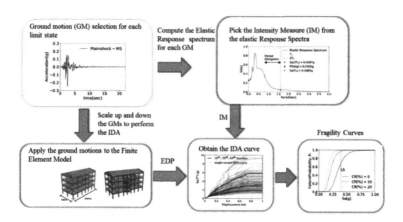

Fig. 9.14 Finite element Model in Seismostruct

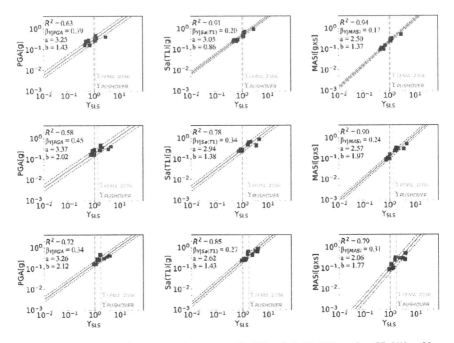

Fig. 9.15 EDP-IM for different corrosion rates **a** CR [%] = 0; **b** CR [%] = 10; **c** CR [%] = 20

9.7.3 Results and Discussion

The results obtained from non-linear dynamic analyses for all IMs can be seen in Fig. 9.15. The inter-storey displacement from FEMA 356 (2000) equal to 2% for the severe limit state is also included in the power interpolation (light blue line) as a measure of the safety level indication of technical codes when the RC structure is exposed to different levels of corrosion. MASI appears to be the most efficient and proficient seismic IM herein examined. The least standard deviation of the residuals describes its high efficiency, which could probably lie in the relevant inelastic effects of the higher modes included in the range of period T_1—$2T_1$ that allows to capture the degree of non-linearity of the structural response. The interested reader is advised to also check Luco and Cornell (2001), who consider a similar IM. By contrast, PGA demonstrated the largest dispersion of the results, and the lowest effectiveness related to Pearson's coefficient. The last observation could be found in the lack of correlation with both the structural parameters and the inelastic damage. Conversely, $S_a(T_1)$ still appears to be efficient, even if less practical than PGA.

Moreover, the value of the inter-storey drift from technical codes seems to overpredict the safety levels for existing structures over their lifetime, as can be clearly seen for corrosion rates between 10 and 20%. The effects of corrosion are increasing the demand in terms of inter-storey displacement while decreasing the capacity of the structure taken from non-linear static analyses. This is primarily due to the reduction

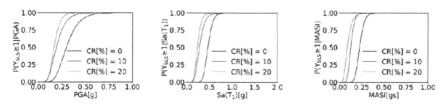

Fig. 9.16 Seismic fragility curves for different corrosion rates and IMs (SD)

of the main mechanical properties of both concrete and steel reinforcement, which affects the local and the global response of the building itself. The power interpolation provides a real perspective of the lack of information in seismic codes when degradation and damage factors over time alter the pristine condition of RC structures. Limit thresholds in current technical codes are completely defined by means of inter-storey drift ratio (i.e. taken equal to 2% for Severe Damage) for as-pristine structures. However, such codes do not account for the effects of deterioration over time. Environmental factors (i.e. corrosion), in fact, could cause additional damage and therefore lead to an overestimation of the actual response of RC structures. As a result, the fragility curves are presented considering the limit proposed in Di Sarno and Pugliese (2020) and mentioned previously in Table 9.7.

Figure 9.16 shows that corrosion has a significant impact on the seismic capacity of the RC building for the selected ground motion excitations. The occurrence of the SD appears for higher values of scaled records which implies a more evident impact of corrosion. The range of values that would cause the exceedance of the limit state decreased, while the damage probabilities for all IMs increased, in comparison with the structures subjected to earthquake excitations in the pristine condition (i.e. CR[%] = 0). The probability of failure equal to 100% is moving left from 0.6 to 0.25 g for PGA, from 0.55 g to 0.4 g for $S_a(T_1)$ and from 0.35 g to 0.15 g for MASI. The last observations demonstrate that corrosion is significantly affecting the safety level for the structure itself.

Figure 9.17 shows the dramatic increment of the seismic vulnerability for all the examined IMs. The corrosion impact for the RC building, subjected to the same ground motion as for the pristine case, forces the structure to undergo higher inter-storey displacements to such an extent that even smaller earthquakes could cause its failure and collapse. $S_a(T_1)$ exhibited the highest values of the decay of the seismic

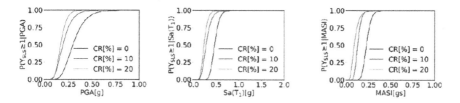

Fig. 9.17 Seismic fragility curves for different corrosion rates and IMs (SD)

vulnerability with the increase of the corrosion rate (e.g. 75% and 85% for CRs equal to 10% and 20% respectively). Conversely, PGA demonstrated the lowest values of the failure probability, which were less than 50% and 65% for the investigated corrosion rates. MASI produced similar results to $S_a(T_1)$ (e.g. an increase of the probability of exceedance equal to 75% and 85% respectively). The similar trend for $S_a(T_1)$ and MASI may be justified by the small differences in the characterization of the power interpolation efficiency of the seismic IMs (Fig. 9.15).

9.8 Conclusions

Fragility functions are an effective assessment tool that can be used by engineers, analysts and policy-makers to determine pre-emptive measures prior to and planning the response in the aftermath of an earthquake. This chapter identified and explained the major steps to undertake fragility analysis and presented two case studies that further emphasise the methodology. Analytical fragility functions are the most widely employed form due to recent advances in analytical methods and the dearth of actual performance data. Different analysis methods can be employed, since they are all robust; however, attention should be given regarding which method is the most applicable under the conditions of the application. For instance, it was demonstrated in Sect. 9.4.2 that the cloud method can yield the same results for a region of interest in comparison with computationally demanding methods, such as IDA and MSA. Additionally, solution measures can vary depending on the analysis method and the types of uncertainties considered.

Two analytical CSs addressed the seismic reliability of a steel and RC building accounting for aftershock and corrosion effects using the IDA method. The approximate closed-form solutions of lognormal distribution are widely accepted for different structural configurations, and thus they were adopted in the CSs. Finally, the most common indicators of failure, as well as efficient and sufficient IMs were identified and demonstrated through the CSs. For example, the spectral acceleration (S_a) is the most predominant measure for low- and mid-rise buildings; however, in the CS#2, it was demonstrated that MASI was the most proficient IM compared to S_a and PGA. MASI is able to account for the structural elongation due to inelasticity and has also been proved efficient for high-rise buildings.

Overall CS#1 demonstrated that RC buildings are likely to lose most of their capacity due to aftershocks when the structure is classified at or beyond the immediate occupancy damage level. In that case, the code-assessment process should account for the existing damage either through the reduction of performance criteria threshold values or an increase of seismic action. Moreover, CS#2 illustrated that corrosion can significantly affect the resistance of RC buildings regardless of the IM considered. Nevertheless, EN 1998–3 (2004) is not always applicable when the structure has experienced corrosion, which urges a code amendment in the next revised versions.

9.9 Future Challenges

Although a solid ground has been formed in the framework of fragility analysis of common buildings, there is still research to be done in the fragility analysis of structures in the following aspects:

- Consideration of ageing effects accounting for the time-dependence of the corrosion phenomenon (i.e. initiation and propagation). This will allow the fragility derivation based on the lifetime and not on the corrosion rates.
- Examination of aftershock events considering near- and far-field conditions as well as soil structure interaction.
- Life-cycle assessment of structures accounting for different uncertainties (e.g. corrosion and aftershock effects using numerical analysis), since the cost of a posteriori interventions can overtop the cost of prudent design.
- Estimation of seismic fragility of non-structural components inside critical facilities and integration of different failure modes that can contribute in the overall risk of the system. The integration is more critical for special structures (e.g. in healthcare facilities or industrial plants).

References

Ancheta TD et al (2013) PEER NGA-West2 Database, PEER Report

Andisheh K, Scott A, Palermo A (2019) Experimental evaluation of the residual compression strength and ultimate strain of chloride corrosion-induced damaged concrete. Struct Conc. https://doi.org/10.1002/suco.201800108

ASCE 41-06 (2005) Seismic rehabilitation of existing buildings, American Society of Civil Engineering. Reston, Virginia

ASCE 41-13 (2013) Seismic evaluation and retrofit of existing buildings. American Society of Civil Engineers, Reston, VA.https://doi.org/10.1061/9780784412855

Bakalis K, Vamvatsikos D (2018) Seismic fragility functions via nonlinear response history analysis. J Struct Eng (United States). https://doi.org/10.1061/(ASCE)ST.1943-541X.0002141

Baker JW (2015) Efficient analytical fragility function fitting using dynamic structural analysis. Earthq Spect. https://doi.org/10.1193/021113EQS025M

Biskinis D, Fardis MN (2009) Upgrading of resistance and cyclic deformation capacity of deficient concrete columns. Geotechn, Geologic Earthq Eng. https://doi.org/10.1007/978-90-481-2681-1_15

Calvi GM et al (2006) Development of seismic vulnerability assessment methodologies over the past 30 years'. J Earthq Tech, (September)

Coronelli D, Gambarova P (2004) Structural assessment of corroded reinforced concrete beams: Modeling guidelines. J Struct Eng. https://doi.org/10.1061/(ASCE)0733-9445(2004)130:8(1214)

De Risi R, Di Sarno L, Paolacci F (2017) Probabilistic seismic performance assessment of an existing RC bridge with portal-frame piers designed for gravity loads only. Eng Struct 145:348–367. https://doi.org/10.1016/j.engstruct.2017.04.053

Di Sarno L et al (2017) Experimental response of an existing RC bridge with smooth bars and preliminary numerical simulations. Eng Struct. https://doi.org/10.1016/j.engstruct.2017.01.052

Di Sarno L, Pugliese F (2020) Numerical evaluation of the seismic performance of existing rein-forced concrete buildings with corroded smooth rebars. Bulletin Earth Eng 18(9):4227–4273. https://doi.org/10.1007/s10518-020-00854-8

Di Sarno L, Karagiannakis G (2020) On the seismic fragility of pipe rack—piping systems considering soil–structure interaction. Bull Earthq Eng. https://doi.org/10.1007/s10518-020-007 97-0

Di Sarno L, Pugliese F (2020) Seismic fragility of existing RC buildings with corroded bars under earthquake sequences. Soil Dyn Earthq Eng 134:106169. https://doi.org/10.1016/j.soildyn.2020. 106169

Dolsek M (2009) Incremental dynamic analysis with consideration of modeling uncertainties. Earthq Eng Struct Dyn. https://doi.org/10.1002/eqe.869

Elnashai AS, Di Sarno L (2015) Fundamentals of earthquake engineering: from source to fragility. Wiley, Second. https://doi.org/10.1002/9780470024867.fmatter/pdf

Elnashai AS, Borzi B, Vlachos S (2004) Deformation-based vulnerability functions for RC bridges. Struct Eng Mech. https://doi.org/10.12989/sem.2004.17.2.215

EN 1998–3 (2004) Eurocode 8: design of structures for earthquake resistance—Part 3: assessment and retrofitting of buildings. CEN, pp 1–97

EN1998–1 (2004) Eurocode 8: Design of structures for earthquake resistance—Part 1 : general rules, seismic actions and rules for buildings, European Committee for Standardization [Authority: The European Union per Regulation 305/2011, Directive 98/34/EC, Directive 2004/18/EC]

FEMA 356 (2000) Prestandard and commentary for the seimic rehabilitation of buildings, rehabilitation requirements

Filippou FC, Popov EP, Bertero VV (1983) Effects of bond deterioration on hysteretic behaviour of reinforced concrete joints. Report to the National Science Foundation, Earthquake Engineering Research Center

Gupta A, Krawinkler H (1999) Seismic demands for performance evaluation of steel moment resisting frame structures. John A, Blume Earthquake Engineering Center Technical Report Series

HAZUS-MH MR5 (2010) Multi-hazard loss estimation methodology earthquake model, Depart-ment of Homeland Security, Federal Emergency Management Agency (FEMA), Mitigation Division Washington, D.C

Ibarra LF, Krawinkler H (2005) Global collapse of frame structures under seismic excitations, evaluation

Iervolino I, Fabbrocino G, Manfredi G (2004) Fragility of standard industrial structures by a response surface based method. J Earthq Eng. https://doi.org/10.1142/S1363246904001717

Iervolino I, Galasso C, Cosenza E (2010) REXEL: Computer aided record selection for code-based seismic structural analysis. Bull Earthq Eng 8(2):339–362. https://doi.org/10.1007/s10518-009-9146-1

Imperatore S, Rinaldi Z, Drago C (2017) Degradation relationships for the mechanical properties of corroded steel rebars. Const Build Mat. https://doi.org/10.1016/j.conbuildmat.2017.04.209

Jalayer F ((2003) Direct probabilistic seismic analysis: implementing non-linear dynamic assess-ments, Department of Civil and Environmental Engineering, Ph.D thesis

Jalayer F et al (2017) Analytical fragility assessment using unscaled ground motion records. Earthq Eng Struct Dyn. https://doi.org/10.1002/eqe.2922

Jalayer F, Cornell CA (2009) Alternative non-linear demand estimation methods for probability-based seismic assessments. Earthq Eng Struct Dyn. https://doi.org/10.1002/eqe.876

Jeon JS et al (2015) Framework of aftershock fragility assessment-case studies: older California reinforced concrete building frames. Earthq Eng Struct Dyn. https://doi.org/10.1002/eqe.2599

K-NET (2019) National Research Institute for Earth Science and Disaster Resilience. NIED K-NET, KiK-net, National Research Institute for Earth Science and Disaster Resilience

Kwon OS, Elnashai A (2006) The effect of material and ground motion uncertainty on the seismic vulnerability curves of RC structure. Eng Struct. https://doi.org/10.1016/j.engstruct.2005.07.010

Kwon OS, Elnashai AS (2010) Fragility analysis of a highway over-crossing bridge with consideration of soil-structure interactions. Struct Infrastruct Eng. https://doi.org/10.1080/157324708 02663870

Liberatore L, Decanini L (2011) Effect of infills on the seismic response of high-rise RC buildings designed as bare according to Eurocode 8. Ingegne-Ria Sismica 3:7–23

Luco N, Cornell CA (2001) Structure-specific scalar intensity measures for near-source and ordinary earthquake ground motions. Earthq Spectra, Doi 10(1193/1):2723158

Luzi L, Pacor F, Puglia R (2019) Italian Accelerometric Archive v3.0., Istituto Nazionale di Geofisica e Vulcanologia, Dipartimento della Protezione Civile Nazionale

Mackie K, Stojadinovic B (2003) Seismic demands for performance-based design of bridges, PEER Report 2003/16 Pacific Earthquake Engineering Research Center College of Engineering University of California Berkeley

Mackie KR, Stojadinovic B (2005) Comparison of incremental dynamic, cloud, and stripe methods for computing probabilistic seismic demand models. In: Proceedings of the structures congress and exposition. https://doi.org/10.1061/40753(171)184

Mazzoni S et al. (2006) OpenSees command language manual, Pacific Earthquake Engineering Research Center

Menegotto M, Pinto P (1973) Method of analysis for cyclically loaded rein-forced concrete plane frames including changes in geometryand non-elastic behavior of elements under combined normal force and bending. In: Proceedings of IABSE symposium on resistance and ultimate deformability of structures acted on by well-defined repeated loads, International Assoc. of Bridge Struct Eng 13:15–22

Miano A et al (2018) Cloud to IDA: efficient fragility assessment with limited scaling. Earth Eng Struct Dyn. https://doi.org/10.1002/eqe.3009

Mohammad Noh N et al (2017) Modelling of masonry infilled RC frames subjected to cyclic loads: state of the art review and modelling with OpenSees. Eng Struct. https://doi.org/10.1016/j.engstruct.2017.07.002

NTC (2018) "Norme Tecniche per le costruzioni", DM Infrastructure, 14 January. (in Italian)

Padgett JE, Nielson BG, DesRoches R (2008) Selection of optimal intensity measures in probabilistic seismic demand models of highway bridge portfolios. Earthq Eng Struct Dyn. https://doi.org/10.1002/eqe.782

Panchireddi B, Ghosh J (2019) Cumulative vulnerability assessment of highway bridges considering corrosion deterioration and repeated earthquake events. Bull Earthq Eng. https://doi.org/10.1007/s10518-018-0509-3

Pinto PE, Franchin P (2010) Issues in the upgrade of Italian highway structures'. J Earthq Eng. https://doi.org/10.1080/13632461003649970

Pitilakis K et al (2014) SYNER-G: typology definition and fragility functions for physical elements at seismic risk. Springer Science & Business Media, Dordrecht. https://doi.org/10.1007/978-94-007-7872-6

Porter KA, Beck JL, Shaikhutdinov RV (2002) Investigation of sensitivity of building loss estimates to major uncertain variables for the Van Nuys testbed investigation of sensitivity of building loss estimates to major uncertain variables for the Van Nuys testbed. PEER Center Report. Doi 10(1193/1):1516201

Razvi SR, Saatcioglu M (1994) Strength and deformability of confined high-strength concrete columns. ACI Struct J. https://doi.org/10.14359/1499

Schultz MT et al (2010) ERDC SR-10-1 Beyond the factor of safety: developing fragility curves to characterize system reliability

SeismoSoft (2019) A computer program for static and dynamic nonlinear analysis of framed structures. https://www.seismosoft.com

Shome N, Cornell CA (1999) Probabilistic seismic demand analysis of nonlinear structures, Report No. RMS-35. Reliability of Marine Structures Program, Department of Civil and Environmental Engineering, Stanford University

Vamvatsikos D, Cornell CA (2002) Incremental dynamic analysis. Earthq Eng Struct Dyn 31(3):491–514. https://doi.org/10.1002/eqe.141

Vamvatsikos D, Cornell CA (2005) Developing efficient scalar and vector intensity measures for IDA capacity estimation by incorporating elastic spectral shape information. Earthq Eng Struct Dyn. https://doi.org/10.1002/eqe.496

Vamvatsikos D, Fragiadakis M (2010) Incremental dynamic analysis for estimating seismic performance sensitivity and uncertainty. Earthq Eng Struct Dyn 39(2):141–163. https://doi.org/10.1002/eqe.935

Vamvatsikos D, Cornell CA (2004) Applied incremental dynamic analysis. Earthq Spect. Doi 10(1193/1):1737737

Vecchio FJ, Collins MP (1986) Modified compression-field theory for reinforced concrete elements subjected to shear. J Am Conc Inst. https://doi.org/10.14359/10416

Wang XH, Liu XL (2008) Modeling the flexural carrying capacity of corroded RC beam. J Shanghai Jiaotong Univ (Science). https://doi.org/10.1007/s12204-008-0129-1

Chapter 10
Earthquake Physical Risk/Loss Assessment Models and Applications: A Case Study on Content Loss Modeling Conditioned on Building Damage

S. Akkar

Abstract This paper presents a novel approach to develop content fragility conditioned on building damage for contents used in residential buildings in Turkey. The approach combines the building damage state probabilities with the content damage probabilities conditioned on building damage states to develop the content fragilities. The paper first presents the procedure and then addresses the epistemic uncertainty in building and content fragilities to show their effects on the content vulnerability. The approach also accounts for the expert opinion differences in the content replacement cost ratios (consequence functions) as part of the epistemic uncertainty. Monte Carlo sampling is used to consider the epistemic uncertainty in each model component contributing to the content vulnerability. A sample case study is presented at the end of the paper to show the implementation of the developed content fragilities by calculating the average annual loss ratio (AALR) distribution of residential content loss over the mainland Turkey.

10.1 Introduction

Porter (2019) defines probabilistic seismic risk as the relationship between the rate (or probability) of an undesirable outcome (e.g., structural collapse, business interruption) and a measurable metric (e.g., money) to quantify the undesirable outcome. The components involved in risk assessment are:

(a) the target exposure: a single asset or an inventory located at a specific site or a region,

(b) the seismic hazard: explaining the exceedance frequency of the ground-motion intensity measure (GMIM) used in defining the conditional probability of the undesirable outcome and,

S. Akkar (✉)
Department of Earthquake Engineering, Kandilli Observatory and Earthquake Research Institute, Boğaziçi University, Istanbul 34684, Turkey
e-mail: sinan.akkar@boun.edu.tr

© The Author(s) 2021
S. Akkar et al. (eds.), *Advances in Assessment and Modeling of Earthquake Loss*,
Springer Tracts in Civil Engineering,
https://doi.org/10.1007/978-3-030-68813-4_10

(c)   the fragility function: describing the occurrence probability of the undesirable
      outcome conditioned on the GMIM utilized in quantifying the seismic hazard.

When fragility functions conditioned on the ground-motion metric are combined
with the consequence models, we quantify the loss (repair costs, loss of functionality).
The functions measuring the loss in terms of GMIM is referred to as vulnerability
functions or vulnerability models. The above terminology can be found in most of
the modern seismic risk assessment text books (e.g., McGuire 2004).

One of the probabilistic risk metrics used in insurance and reinsurance is the
average annual loss (AAL) that measures the expected annual loss over a predefined range of GMIM with different exceedance frequencies (Eq. 10.1). If, for
example, the loss is measured in terms of repair cost, AAL is the average money
that would be spent every year to repair the asset. To this end, $y(s)$ in Eq. (10.1) is the
vulnerability function in terms of GMIM, s, and the derivative of the seismic hazard
curve $G(s)$ represents the annual probability producing exactly s. The negative sign
accounts for the negative slope of $G(s)$ at s since the hazard curve slopes down to the
right at all values of s indicating lower exceedance frequency of higher shaking.

$$AAL = \int_{s=0}^{\infty} y(s) \frac{-dG(s)}{ds} ds \qquad (10.1)$$

Equation (10.2) shows the vulnerability function computation, as described in
the HAZUS report (FEMA 2003), by considering the fragility functions providing
damage probabilities at different damage states ($\Pr(DS = ds_i)$, $i = 1,\ldots,n$; n is the
total number of damage states). The variable RC^{ds_i} is the replacement cost corresponding to $DS = ds^i$. The replacement costs are the monetary losses, representatives
of different damage levels and are, therefore, called as consequence functions (or
models). The vulnerability function developed by Eq. (10.2) is called as compound
loss function in the HAZUS report since it accounts for all possible damage states,
proportional to their occurrence probabilities, that the asset can experience during
an earthquake.

$$y(s) = RC^{ds_1} \cdot \Pr(DS = ds_1) + \ldots + RC^{ds_i}$$
$$\cdot \Pr(DS = ds_i); \ i = 1 \ldots n \qquad (10.2)$$

In essence, the integral expression in Eq. (10.1) computes the expected annual
loss of an asset by considering a range of GMIM, s, that are likely to occur at the site
with different annual probabilities. If the consequence model used in Eq. (10.2) is
dimensionless (in terms of replacement cost ratio), the resulting loss by Eq. (10.1) is
called as average annual loss ratio (AALR); favored more by the insurance industry.

The accuracy of the predicted loss (in this case AAL or AALR) is confined to
the reliable seismic hazard and vulnerability models hence the consistent fragility
and consequence functions. This fact brings forward the modeling uncertainty (epistemic uncertainty) in these components that is addressed in a variety of scientific
publications. A fairly ample review, in this respect, can be found in FEMA P-58

(ATC 2018). The lack of knowledge, insufficient data and subsequent assumptions as well as interpretations about the model behavior are the main sources of epistemic uncertainty.

This article presents a case study on modeling the uncertainty in content vulnerability functions for residential buildings in Turkey and its progressive influence on the loss computations. Since vulnerability functions are composed of fragility and consequence models (Eq. (10.2)), the progressive influence of the epistemic uncertainty is discussed by considering the interaction between these two modeling elements. The contents considered here are poorly anchored or unanchored house utensils as well as furniture and electronic equipment frequently used in the residential dwellings. Their fragility modeling presented here is conditioned on the different levels of building damage that would lead to more accountable loss predictions.

The paper starts by describing the development of content fragilities conditioned on the building damage that is followed by the development of consequence functions for different modes of content damage. The associated epistemic uncertainty in the conditional content fragilities and consequence functions are progressed to observe their influence on the content vulnerability model. To illustrate the implementation of the discussions, the last part in the paper integrates the conditional vulnerability model together with the most recent national seismic hazard maps to compute the distribution of residential building content AALR for entire Turkey.

10.2 Development of Content Fragilities Conditioned on Building Damage

10.2.1 Review of Some Benchmark Documents

ATC-13 (1985), the HAZUS technical report (FEMA 2003) and FEMA P-58 (ATC 2018) are comprehensive documents proposing methodologies to predict content damage loss for different structural facilities due to ground shaking. ATC-13 (1985), which can be considered as one of the pioneer guidelines in earthquake loss assessment provides content (equipment) fragilities as functions of Modified Mercalli Intenisty (MMI) GMIM. The fragilities are developed by the expert opinions of a group of earthquake engineering specialists. The expert-based content fragilities by ATC-13 (1985) rely on the assumptions of

(a) regular building construction quality,
(b) structures on firm soil,
(c) merely ground shaking without damage aggravation due to collateral hazard (e.g., fire, fault rupture and inundation) and,
(d) the content is at the ground level and unanchored.

Assuming that the content damage is floor acceleration sensitive, the HAZUS technical report (FEMA 2003) provides an empirical formulation for content loss that depends on

(a)   structural damage states,
(b)   building-type sensitive content replacement value and,
(c)   the probability of building being in non-structural acceleration sensitive damage state.

FEMA P-58 (ATC 2018) apriori assumes that the contents are sensitive to peak floor acceleration and velocity, and provides building-type dependent content fragilities as well as normative content quantitates to predict the content loss under ground shaking. The content loss assessment tools provided in the FEMA P-58 and HAZUS documents are more comprehensive than the one in ATC-13 but it seems that they are more suitable to assess a specific single asset as long as the content loss is of concern. Naturally, all three documents establish their methodologies considering the structural typologies and construction quality in the United States. In fact, ATC-13 (1985) was prepared for earthquake loss in California.

10.2.2   Theoretical Background

The content fragilities developed in this section are also conditioned on the building damage as in the case of the HAZUS technical report but their underlying theory utilizes total probability theorem as partially illustrated in Fig. 10.1. In other words, the content damage probability is decomposed into different states $(ds_1^{cnt},\ldots, ds_5^{cnt})$ that are jointly distributed with different building damage states. As presented in

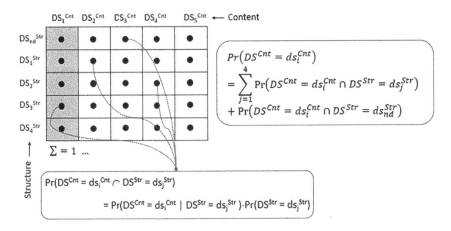

Fig. 10.1 Schematic diagram showing the dependency (interaction) of content and structural damage and its representation in probabilistic terms

Fig. 10.1, the probability that the content is at the damage state $DS^{Cnt} = ds_i^{cnt}$ is the sum of the joint probabilities of the content damage $DS^{Cnt} = ds_i^{cnt}$ and all building damage states, $DS^{Str} = ds_j^{str}$, $j = 1,\ldots,n_{str}$, as well as the likelihood of building being not damaged, $DS^{Str} = ds_{nd}^{str}$.

Equation (10.3) defines the same content damage probability in terms of conditional probabilities where $Pr\left(DS^{Cnt} = ds_i^{Cnt} | DS^{Str} = ds_j^{Str}\right)$ is the ith state content damage probability conditioned on the jth state building damage and $Pr\left(DS^{Cnt} = ds_i^{Cnt} | DS^{Str} = ds_{nd}^{Str}\right)$ is the ith state content damage probability conditioned on undamaged building state. The terms $Pr\left(DS^{Str} = ds_j^{Str}\right)$ and $Pr\left(DS^{Str} = ds_{nd}^{Str}\right)$ refer to the jth damage state and no damage state probabilities of the building, respectively.

$$
\begin{aligned}
\mathrm{Pr}(DS^{Cnt} = ds_i^{Cnt}) = &\sum_{j=1}^{4} \mathrm{Pr}(DS^{Cnt} = ds_i^{Cnt} | DS^{Str} = ds_j^{Str}) \cdot \mathrm{Pr}(DS^{Str} = ds_j^{Str}) \\
& + \mathrm{Pr}(DS^{Cnt} = ds_i^{Cnt} | DS^{Str} = ds_{nd}^{Str})
\end{aligned}
\tag{10.3}
$$

The difficulty in the implementation of Eq. (10.3) is the unknown conditional probabilities (i.e., $Pr\left(DS^{Cnt} = ds_i^{Cnt} | DS^{Str} = ds_j^{Str}\right)$). The other probabilities, that is, the probabilities of building being in a damage state ds_j^{Str} or being in no damage state $DS^{Str} = ds_{nd}^{Str}$ can be determined from the building fragilities. The way to surmount the unknown conditional probabilities can be the utilization of Monte Carlo sampling technique as explained in the next paragraph but before that Fig. 10.2 shows the discrete building and content damage states that are used in their fragility functions and the corresponding descriptions of the content damage states for clarity.

Figure 10.3 displays the concept in proposed approach while generating the content damage probabilities conditioned on the building damage. When the building does not suffer any structural damage, the content is most likely to be slightly damaged for not being properly anchored on the floor. Consequently, the building suffering from light, moderate to severe and very severe (collapse or almost collapse)

Building

- DS$_1^{Str}$: **Light** structural damage
- DS$_2^{Str}$: **Moderate** structural damage
- DS$_3^{Str}$: **Severe** structural damage
- DS$_4^{Str}$: **Very severe** (almost collapse) structural damage

Content

- DS$_1^{Cnt}$: **Slight** (limited localized) content damage not requiring repair
- DS$_2^{Cnt}$: **Light** (significant localized) content damage generally not requiring repair
- DS$_3^{Cnt}$: **Moderate** (significant localized) content damage warranting repair
- DS$_4^{Cnt}$: **Heavy** (extensive) content damage requiring major repairs
- DS$_5^{Cnt}$: **Very heavy** (widespread) content damage either demolished or repaired

Fig. 10.2 Descriptions of building and content damage states

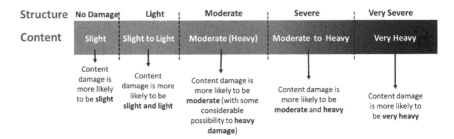

Fig. 10.3 The concept used while generating the content damage probabilities conditioned on building damage

damage states result in slight to very heavy content damages associated with different probabilities. The implementation of this conceptual approach to sample conditional content damage probabilities are illustrated for three different cases in Figs. 10.4, 10.5 and 10.6 representing, respectively, no structural damage, moderate structural damage and very heavy structural damage states of a residential building.

Figure 10.4 indicates that the content damage probabilities are assumed to be represented by three different damage states (slight, light and moderate) when the building sheltering the content does not suffer any structural damage (i.e., when only a limited nonstructural damage is observed in the architectural and mechanical/electrical building components). Note that the approach primarily rates the slight content damage, which is followed by the light and moderate damages associated

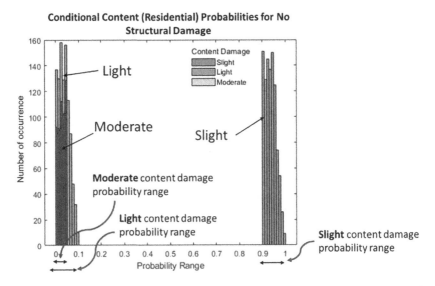

Fig. 10.4 Distribution of content damage probabilities conditioned on no structure damage in the building

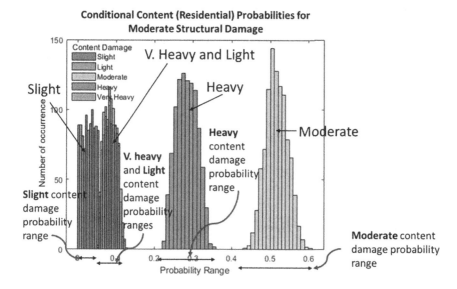

Fig. 10.5 Distribution of content damage probabilities conditioned on moderate structural damage in the building

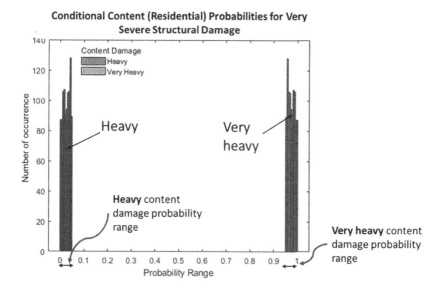

Fig. 10.6 Distribution of content damage probabilities conditioned on very severe structural damage in the building

with very small probabilities. When the building suffers from moderate structural damage (Fig. 10.5), the content damage probabilities are represented by all five states and the proposed approach apriori prefers moderate content damage and then rates the occurrence probability of heavy content damage more than the other three damage states (slight, light and very heavy). Note that the likelihood of very heavy content damage is more than the slight content damage prorating the existence of fragile content in the residential buildings. Upon very severe structural damage (Fig. 10.6), the approach almost exclusively favors very have content damage, practically advocating its full replacement. As a final remark, the sampling should be tailored such that given building damage state, the assigned content damage probabilities should sum up to unity at every sampling.

10.2.3 Case Studies on Developed Content Fragilities

Figures 10.7 and 10.8 show a case on the implementation of the proposed approach. The case study develops content fragilities for a mid-rise reinforced concrete (RC) moment-resisting frame (MRF) residential building, representative of its class, in Turkey. The building damage state probabilities presented in Fig. 10.7 are generated from the fragility study conducted for the Turkish Catastrophe Insurance Pool (TCIP) to revise the compulsory earthquake premiums of residential buildings in Turkey (TDV 2018). The fragility study compiled the reliable fragility models that are representatives of residential buildings in Turkey to develop homogenized fragility models in terms of construction date (No-code -pre1976-, Low-code -1976 to 2000- and High-code -post2000) and story number (Low-rise $-\leq$ 3-story, Mid-rise -4 to 9

Fig. 10.7 Populated damage probabilities at different states for mid-rise (4 to 9 story) reinforced concrete residential building class in Turkey built after 2000 (considered as high-code)

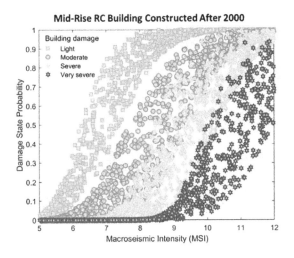

Fig. 10.8 Content damage probabilities at different states for mid-rise, high-code residential building class (see Fig. 10.7) in Turkey

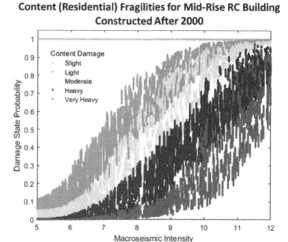

story, and High-rise –9 to 19 story). The fragilities are developed as functions of modified Mercalli Intensity (MMI) scale because some of the fragility models compiled during the fragility study were developed from on-site damage observations after the earthquakes.

The sampled building damage probabilities presented in Fig. 10.7 account for the variability at each damage state due to:

a.   uncertainty in the damage state threshold by different studies,
b.   variability in the modeling aspects of the buildings and,
c.   variability in building response due to intricate nature of earthquake ground-motion records.

For this reason, the Monte Carlo sampling results in bands of probabilities given a specific MMI value. The upper and lower end of the bands, hence the damage probabilities, overlap each other due to excessive variability in the fragilities at each damage state.

Figure 10.8 shows the content fragilities for the same residential building class after implementing Eq. (10.3). The conditional content probabilities conditioned on different building damage states and the building damage probabilities are populated as described in the above paragraphs. They are presented in Figs. 10.4, 10.5, 10.6 and 10.7. To this end, the resulting content fragilities account for the model uncertainty due to building response and content damageability associated with differences in the residential equipment, their locations, placements and etc. That's why the overlapping of damage probabilities at different content damage states are increased with respect to those presented for the buildings (Fig. 10.7).

Figure 10.9 shows the same content fragilities for no-code, mid-rise MRF residential buildings. Hence, the building damage probabilities are revised in these calculations to account for no-code mid-rise building performance under seismic action.

Fig. 10.9 Content damage probabilities at different states for mid-rise, no-code residential building (see Fig. 10.7) in Turkey

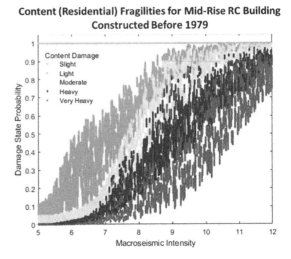

The conditional content damage probabilities are also repopulated by considering the change in the building typology. The comparisons between Figs. 10.8 and 10.9 underline the significance of building performance in the content fragility. Consequently, the contents in low-code and old residential buildings are more susceptible to damage with respect to those in new and high-code buildings.

As for the last discussion in this section, Fig. 10.10 shows the content damage fragilities provided by ATC-13 (1985) for California under the assumptions presented in the beginning of this section. The damage state definitions in this study and the ATC-13 document is the same. ATC-13 provides damage probability matrices

Fig. 10.10 Content damage fragilities provide in ATC-13 (1985) for residential RC buildings in California

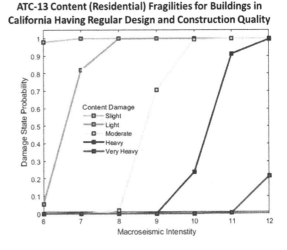

(DPMs) for each target facility class (one of them is the content for residential buildings) after the evaluation of three-round questionnaires. The DPMs for the residential content are converted to fragilities as presented in Fig. 10.10.

The comparisons of content fragilities provided in this study and those given in the ATC-13 document are in agreement to a limited extent. There are major differences in terms of the uncertainty in content damageability predictions by each approach. ATC-13 defines a unique content damage probability at each damage state given a specific MMI value. The content fragilities developed in this study yield a range of damage probability at each content damage state by considering the progression of the epistemic uncertainty in building response as well as content damageability. As depicted by Figs. 10.9 and 10.10, the slight content damage is similarly predicted by the two studies but the ATC-13 probabilities for the rest of the damage states seem to be closer to the lower bound fragility predictions of this study. Hence, upon the use of these two different fragility sets in a probabilistic risk assessment study, one may obtain completely different loss pictures. This fact brings forward the importance of how epistemic uncertainties are handled in a given methodology as well as the country-based differences in loss assessment.

10.3   Content Consequence Model

The replacement cost ratios for content is modeled from expert opinions of risk engineers working in the field of insurance. The experts filled a questionnaire about the replacement cost ratios of a variety of residential equipment (content) subjected to various levels of damage (slight to very heavy as defined in Fig. 10.2). The expert opinions were pooled under a database and several group meetings were held with the experts to reach a common agreement for removing the outlier judgements. The improved expert opinion database was then used to compute the means and the standard deviations of the replacement cost ratios for the predetermined damage states. Given a damage state, the dispersion about its mean replacement cost ratio was used to establish an interval defining the lower and upper boundaries of the replacement cost ratio. This interval accounts for the epistemic uncertainty in the insurance-based replacement cost of the subject content for that specific damage state due to differences in the expert considerations. Figure 10.11 depicts the resulting content model for all five states of the content damage.

The Monte Carlo simulations were utilized to randomly sample the model uncertainty in the content replacement cost ratios. Figure 10.12 displays the randomly sampled replacement cost ratio distributions at each damage state. The replacement cost ratio sampling assumed uniform distribution for this case study though it could also have been any other distribution mimicking a more realistic representation of the replacement cost ratio. The sampled replacement cost ratios and the populated content damage probabilities presented in the previous sections were used in Eq. (10.2) to develop the content vulnerability model that is discussed further in the following section.

Fig. 10.11 Content consequence model (replacement cost ratio) used in this study. The likely replacement cost ratios given a specific content damage state are described within the horizontal gray bars. D1: slight, D2: light, D3: moderate, D4: heavy, D5: very heavy content damages

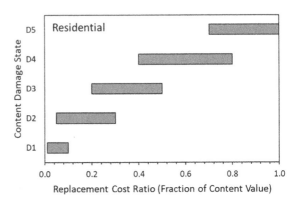

Fig. 10.12 Replacement cost ratio distributions generated from Monte Carlo simulations for each content damage state

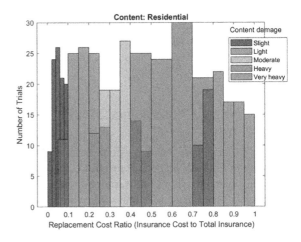

10.4 Vulnerability Model and Country-Wide Content AALR

The sampled content damage probabilities as well as the replacement cost ratios as discussed in the last two sections were input into Eq. (10.2) to develop the compound loses for residential content. As an example, Fig. 10.13 gives the resulting content loss model for high-code, mid-rise RC MRF buildings in Turkey. The gray cloud in this figure illustrates all possible combinations of the sampled content damage probabilities and the replacement cost ratios. The content losses computed from the combinations of damage probabilities and replacement cost ratios depict a very large range, which indicates the progression of epistemic uncertainty from building damage probabilities to content loss conditioned on the building damage as well as the monetary consequences of content damaged at different levels. The median loss (red curve) represents the central trend in the content loss whereas the blue scatters represent the body of the content loss predictions. In other words, the median loss

Fig. 10.13 Content loss (vulnerability) dispersion for mid-rise and high-code RC buildings after combining the content damage probabilities and the consequential replacement cost ratios discussed in the previous sections

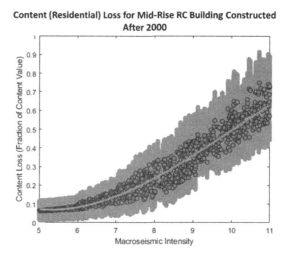

can be interpreted as the most expected (best estimate) content loss trend, whereas the boundaries represented by the body (blue scatters) yield the higher and lower loss values of alternative interpretations in the content damage probabilities as well as the replacement costs. The upper and lower limits of the range (gray cloud) can be interpreted as the extreme content losses that are beyond the empirical data but still have justifiable bases if they are to incur. Note that the increased ground-motion intensity results in higher dispersion in the content loss. This fact advocates a more dominant epistemic uncertainty in the main components of the content vulnerability with increasing ground-motion amplitude.

The last figure in this paper (Fig. 10.14) presents the AALR distribution for residential content in the mid-rise, high-code MRF buildings in Turkey. The median content vulnerability curve as presented in Fig. 10.13 and the soil-condition sensitive

Fig. 10.14 AALR distribution for mainland Turkey for residential content in high-code, mid-rise RC MRFs

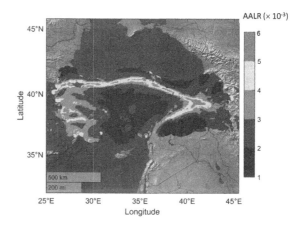

hazard curves developed from the most recent national seismic hazard map (Akkar et al. 2018) were used in Eq. (10.1) for calculations. The site conditions for the entire country were adopted from the overall countrywide geology whereas the PGA and PGV values at different annual exceedance rates by the national seismic hazard maps were converted to MMI via Wald et al. (1999) for estimating the MMI-based hazard. The details of these calculations are discussed in the TDV report (TDV 2018).

The maps suggest an AALR interval ranging between 4×10^{-3} and 6×10^{-3} for the most seismic prone settlements in Turkey (e.g., Istanbul, Izmir, Canakkale, Erzincan, Aydin, Denizli, etc.). This value suggests a yearly basis pure premium of €40 to €60 Euros for residential equipment of €10,000 worth in such cities. The AALR values go down to as much as 1×10^{-3} (i.e., a yearly basis pure premium of €10 for residential equipment of €10,000 worth) in the least seismic regions in the country such as the large portion of the south Eastern Turkey and the central Anatolia. Needless to say the presented numbers are valid for the residential content in the mid-rise, high-code, MRF buildings. They would be scaled up and down depending on the building type, construction period and height as partially discussed in Fig. 10.9.

10.5 Summary and Conclusions

This study proposes a procedure to develop content fragilities conditioned on building damage for loss and risk modeling that can be used in computing metrics relevant to insurance and reinsurance. The conditional content fragilities can account for the epistemic uncertainty in assessing the earthquake induced building damage states as well as the different likelihoods of content damage under different modalities of building damage. These uncertainties are handled via Monte Carlo sampling that enables the risk expert to trace forward or backward the progression of model uncertainty and its effects on the computed loss and risk. The proposed procedure is analytical, and its systematic utilization can result in country-specific vulnerability and risk models. This feature makes the procedure appealing because the current well-organized and state-of-the-art tools in this field seem to be tailored for the construction quality and building classification in the US practice (e.g., ATC-13 1985; FEMA 2003). The systematic efforts for improving this procedure should involve calibrations through comparisons with other approaches as well as sensitivity analyses to understand the behavior of critical components contributing the most to loss and risk assessment results.

Acknowledgements This study is done under the financial supports provided by Turkish Catastrophe Insurance Pool and Turkish Insurance Association via Turkish Earthquake Foundation. The author benefitted significantly from the fruitful discussions with Prof. Mustafa Erdik and Prof. Ufuk Yazgan while establishing the presented content fragility model.

References

Akkar S, Eroğlu Azak T, Çan T, Çeken U, Demircioğlu Tümsa MB, Duman TY, Erdik ÖM, Ergintav S, Kadirioğlu FT, Kalafat D, Kale Ö, Kartal RF, Kekovalı K, Kılıç T, Özalp S, Altuncu Poyraz S, Şeşetyan K, Tekin S, Yakut A, Yılmaz MT, Yücemen MS, Zülfikar Ö (2018) Evolution of seismic hazard maps in Turkey. Bull Earthq Eng 16:3197–3228

Applied Technology Council, ATC (1985) ATC-13 Earthquake damage evaluation data for California. Prepared by C. Rojahn and RL Sharpe, funded by Federal Emergency Management Agency

Applied Technology Council, ATC (2018) FEMA P58 Seismic performance assessment of buildings. Funded by Federal Emergency Management Agency

Federal Emergency Management Agency, FEMA (2003) Multi-hazard loss estimation methodology HAZUS®MH MR4 Technical Manual

McGuire R (2004) Seismic hazard and risk analysis. Earthquake Engineering Research Institute, Oakland, CA

Porter K (2019) A beginner's guide to fragility, vulnerability, and risk. University of Colorado Boulder, p 119. https://spot.colorado.edu/~porterka/Porter-beginners-guide.pdf

Turkiye Deprem Vakfi, TDV (2018) Revision of earthquake premiums in accordance with the revised national seismic hazard maps of Turkey (in Turkish)

Wald JW, Quitariano V, Heaton TH, Kanamori H (1999) Relationships between peak ground acceleration, peak ground velocity, and modified Mercalli intensity in California. Earthq Spect 15:557–564

Chapter 11
Earthquake Catastrophe Risk Modeling, Application to the Insurance Industry: Unknowns and Possible Sources of Bias in Pricing

M. Kohrangi, A. N. Papadopoulos, S. R. Kotha, D. Vamvatsikos, and P. Bazzurro

Abstract Mathematical risk assessment models based on empirical data and supported by the principles of physics and engineering have been used in the insurance industry for more than three decades to support informed decisions for a wide variety of purposes, including insurance and reinsurance pricing. To supplement scarce data from historical events, these models provide loss estimates caused to portfolios of structures by simulated but realistic scenarios of future events with estimated annual rates of occurrence. The reliability of these estimates has evolved steadily from those based on the rather simplistic and, in many aspects, semi-deterministic approaches adopted in the very early days to those of the more recent models underpinned by a larger wealth of data and fully probabilistic methodologies. Despite the unquestionable progress, several modeling decisions and techniques still routinely adopted in commercial models warrant more careful scrutiny because of their potential to cause biased results. In this chapter we will address two such cases that pertain to the risk assessment for earthquakes. With the help of some illustrative but simple applications we will first motivate our concerns with the current state of practice in modeling earthquake occurrence and building vulnerability for portfolio risk assessment. We will then provide recommendations for moving towards a more comprehensive, and arguably superior, approach to earthquake risk modeling that

M. Kohrangi (✉)
RED Risk Engineering + Development, Via Giuseppe Frank No. 38, Pavia, Italy
e-mail: mohsen.kohrangi@redrisk.com

A. N. Papadopoulos
Swiss Seismological Service, ETH Zurich, Switzerland

S. R. Kotha
Univ. Grenoble Alpes, Univ. Savoie Mont Blanc, CNRS, IRD, IFSTTAR, ISTerre, 38000 Grenoble, France

D. Vamvatsikos
School of Civil Engineering, National Technical University of Athens, Athens, Greece

P. Bazzurro
University School for Advanced Studies IUSS Pavia, Pavia, Italy

© The Author(s) 2021   239
S. Akkar et al. (eds.), *Advances in Assessment and Modeling of Earthquake Loss*,
Springer Tracts in Civil Engineering,
https://doi.org/10.1007/978-3-030-68813-4_11

capitalizes on the progress recently made in risk assessment of single buildings. In addition to these two upgrades, which in our opinion are ready for implementation in commercial models, we will also describe an enhancement in ground motion prediction that will certainly be considered in the models of tomorrow but is not yet ready for primetime. These changes are implemented in example applications that highlight their importance for portfolio risk assessment. Special consideration will be given to the potential bias in the Average Annual Loss estimates, which constitutes the foundation of insurance and reinsurance policies' pricing, that may result from the application of the traditional approaches.

11.1 Introduction

Since the early 1990's the use of catastrophe risk models has been adopted in the insurance industry to estimate the likelihood of observing losses in a given period of time due to the occurrence of natural events, such as earthquakes, tropical and extra-tropical cyclones, and floods. Too few and scarcely representative historical loss data were available to support a robust estimation via traditional statistical techniques or expert judgment, especially for large events that were not yet observed in recent historical times. In essence these probabilistic catastrophe risk models, which are based on applying the best available science on the existing data, were used to simulate virtual loss observations to augment scarce or missing real loss observations. Since their advent, these models have been used by insurance/reinsurance companies, rating agencies, hedge funds, catastrophe risk pools, mortgage lending institutions, governments and corporations, among others, for all sort of risk management decisions. One such a decision involves setting the premium of insurance/reinsurance policies for single assets and portfolios of assets. We will dwell on insurance pricing towards the end of this chapter.

All these catastrophe risk models have all the same structure and include *four modules* regardless of which natural event they are meant to address, only the details differ from one peril to another.

Firstly, an *exposure module*, which describes all the characteristics of the assets at risk in the region of interest. These assets usually include all the building inventory and, sometimes, infrastructures. In many insurance applications, however, the exposure is not the portfolio of the entire built environment but simply the specific portfolio of assets to be insured or reinsured. Each asset in the portfolio is traditionally assigned to one of the many different classes of structures (e.g., midrise reinforced concrete frame buildings of the 1970's era) that exhibit a different level of vulnerability to the natural events under consideration.

The second is the *hazard module*, which is conceptually divided into two parts: an event occurrence sub-module, which defines the rate of occurrence of future events in terms of size and location. This sub-module is responsible for producing stochastic catalogs of simulated future events that are statistically consistent (but not identical) to those occurred in the past. The second sub-module deals with the predictions of the effects that each event in the stochastic catalog may cause in the nearby region. The effects may be, for example, ground motion for earthquakes, or wind, storm surge, and precipitation for tropical cyclones. For example, in the case of earthquakes, given the occurrence of an event of given magnitude, M, this sub-module provides the distribution of the intensity measure (IM) for any site at a given distance, R, from the rupture with local soil characteristics, often conventionally expressed in terms of the average shear wave velocity, Vs_{30}, in the top 30 m.

The third is the *vulnerability module*, which contains for all different types of assets the relationships that provide the level of loss ratio (namely a loss expressed as a percentage of the total replacement cost of the asset) and its variability expected for any given intensity level of the effect that the asset may experience (e.g., for Peak Ground Acceleration of 0.05 to 3.0 g in the case of earthquakes or wind speed from 50 to 300 km/h for tropical cyclones) in its lifetime. These relationships that associate an IM of the effect of the event with a loss ratio are called vulnerability functions. Vulnerability functions are often constructed by engineers by convolving two other types of functions: fragility functions and consequence functions. Fragility functions, which are derived for specific damage states (ranging say, from minor damage to collapse), provide the likelihood that an asset will end up in any given damage state should it experience any given level of IM. Consequence functions simply link the damage states to distributions of loss ratios (e.g., a minor damage state may correspond to losses in the 2 to 5% of the replacement cost of the asset). Both vulnerability and fragility functions will be discussed later in the chapter.

The fourth and last is the *Loss Module*, which handles the computations of losses for each event in the hazard module and for each asset in the exposure module by applying the asset-class-specific vulnerability function as specified in the vulnerability module. The losses could be the so-called ground-up losses, which include all the losses that one would need to sustain, for example, to repair or replace all the structural and non-structural components damaged by an event, or the insurance losses. The latter are computed from the former by applying the policy conditions (e.g., deductibles and limits). Note that the monetary losses may in some cases not refer to repair cost but to costs incurred due to downtime, i.e., the time required to make the asset functional again. In other applications, the losses computed are non-monetary and are measured in terms of number of injuries or fatalities caused by the damage and collapse of the assets caused by the event. For monetary losses, the standard outputs of the Loss Module are the Average Annual Loss (AAL) and the so-called Exceedance Probability (EP) Loss Curve. The AAL is the expected loss that the stakeholder can expect to pay, on average, every year over a long period of time. If the stakeholder is, say, the owner of a building the AAL is the amount of money that, on average, is needed every year for fixing the damage caused by natural events. If it is an insurance company, the AAL is the amount of money that, on average, the insurance is expected to pay the insured every year because of the damage caused by natural events. The EP Loss curve provides the annual probability (or rate) that the losses (by single event or aggregated for all the events in a year) will exceed different amounts of monetary values. These losses may refer to a single asset or, in the case of an insurance or reinsurance company, to a portfolio of many assets. These two standard outputs are complementary and fully consistent with each other.

In the rest of the chapter we will discuss some of the caveats of the traditional methodologies applied in commercial catastrophe risk models for assessing earthquake risk. The next section will deal with the event occurrence part of the hazard module and will be followed by a section that will address the shortcoming of the universally adopted approach for deriving vulnerability functions for classes of buildings. We show two applications where we propose enhancements to the current methodologies that we believe are ready to be embraced in the next generation of catastrophe risk models. Then we will discuss the impact on seismic hazard and risk due to the next generation of non-ergodic ground motion prediction equations. This enhancement, which pertains to the second part of the hazard module, is indeed very promising and will certainly revolutionize the ground motion prediction of earthquake models of the near future. However, this new approach is more complex and requires an amount of data that is nowadays only available in very few parts of the world. Therefore, its widespread adoption is still premature and will need additional data and research before it is incorporated in commercial earthquake risk assessment models. Finally, we will show how the potential biases in the loss estimates caused by the caveats of the traditional approaches adopted in the current commercial models may affect insurance pricing for earthquake risk.

11.2 Should Earthquake Sequences be Removed from Seismic Hazard and Risk Assessment Models?

Probabilistic Seismic Hazard Analysis (PSHA) is the methodology universally adopted in the hazard module of all earthquake risk assessment models in use today. In the large majority of PSHA studies and in all those used for loss estimation purposes, the rates of occurrence of events have been derived from historical and instrumental seismicity catalogues that undergo a so-called "declustering" procedure. This procedure involves identifying earthquake clusters (in time and space) and removing from the catalogues all but one event per cluster (typically the one with the largest magnitude). The earthquakes kept are classified as "mainshocks", while those that are discarded are referred to as "foreshocks" or "aftershocks", depending on whether they occurred before or after the mainshock, respectively. In some clusters this procedure removes also "triggered" events that did not break the same part of the fault ruptured by the mainshock as, loosely speaking, "proper" foreshocks and aftershocks do, but adjacent fault segments or, sometimes, different nearby faults altogether. Hereafter, for brevity we will refer to all these removed events simply as aftershocks. The rationale behind this practice is to ensure that the occurrences of earthquakes in the final catalogue are independent events. This convenient property in turn allows the analyst to employ the well-known Poisson process to model the occurrence of future (mainshock-only) seismicity with rates derived from the declustered historical catalogues. The clear advantage of this choice is the mathematical simplicity of the modeling process. On the other hand, one can expect the resulting seismic hazard to be underestimated due to the exclusion of non-mainshock events whose ground motions could exceed significant levels of intensity at the site(s) of interest.

The underestimation of hazard caused by the practice of considering only mainshocks may be less severe in the small minority of advanced PSHA studies that rely on fault (rather than area source) source characterization for which earthquake rates are computed strictly from geologic and geodetic data. In these very few cases, the earthquake activity rate on a fault is computed by balancing the long-term seismic moment build up estimated by geologic or geodetic observations and the seismic moment available for release by future earthquakes. Hence, it may be less critical whether the moment-rate is balanced by only mainshocks of larger magnitude or by entire sequences of events. Enforcing the moment balance using sequences is, however, clearly the preferred approach.

In the early days of PSHA, reducing the phenomenon of seismicity to just mainshock events might have been warranted to facilitate its implementation in the state of practice and also somewhat inevitable given the scarcity of the then-available seismicity data and the limited understanding of the problem. Moreover, the primary application of PSHA at the time was to underpin seismic design codes by defining the ground motion input that a given structure was expected to withstand with a given likelihood within its lifetime. In that context, declustering was largely justified by the appealing, but imprecise, notion that if a structure is designed to withstand the ground motion of the "stronger" (in magnitude) mainshock, it will also be able to survive that of the "weaker" aftershocks. Or, to the very least, if the structure survived the mainshock then the occupants could leave unharmed before the aftershocks would increase the severity of the damage and perhaps destroy it altogether.

This somewhat naïve approach, which was to a certain extent justifiable for that application, has, however, been carried over to earthquake risk modeling. This approach is still routinely followed until the present day, as if all the events in a sequence but the mainshock would cause no additional damage, no further loss of serviceability of buildings, and no additional repair cost. In the insurance industry, for example, to legitimize this simplistic approach it is often reasoned that the effects of aftershocks and triggered events are already implicitly accounted for since the employed vulnerability curves are calibrated to match actual damage data gathered after the entire earthquake sequence has taken place. However, aside from the scarcity of such damage and loss data that only seldom allow the development of empirical vulnerability functions, the validity of the claim that the effects of aftershocks and triggered events can be embedded in conventional vulnerability curves is not only questionable but also lacks a theoretical foundation. Furthermore, even if the claim were true, this approach would imply that using these somewhat "inflated" vulnerability curves would produce overestimated losses in all those cases where no major aftershocks are triggered after the mainshock earthquake.

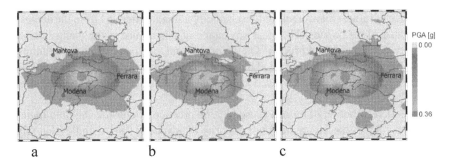

Fig. 11.1 Mean PGA ShakeMaps from INGV for the **a** 20 May M5.9 and **b** 29 May 2012 M5.8 Emilia-Romagna earthquakes. Panel **c** shows a map of the maximum mean PGA values from the two earthquakes

In an attempt to bring to light potential sources of bias, several shortcomings of the current mainshock-only view of seismicity are outlined in the following paragraphs.

11.2.1 Fewer Earthquakes Modeled

The most obvious deficiency—yet hidden in plain sight—is rooted in the main argument for declustering: the expectation that the mainshock earthquake will always be more damaging than its preceding and succeeding lower magnitude events. While generally true, it does not necessarily apply for every structure in the region affected by the seismic cluster. To demonstrate this point, let us look at the potential ground motion intensity that different earthquakes in the same cluster can induce. As an example, Fig. 11.1a and b show the peak ground acceleration (PGA) ShakeMaps[1] provided by the Italian Institute of Geophysics and Volcanology (INGV) for the two M5.9 and M5.8 events in the Emilia-Romagna sequence that struck this region of Italy in May 2012. Figure 11.1c, on the other hand, shows a map of the maximum PGA value at each grid cell from the two earthquakes. It is clear from these plots that there are areas in which the largest expected PGA comes from the first event and areas where the largest expected PGA is induced by the second event. This is arguably a consequence of the distance of these sites to each of the two ruptures, but it could also be partly attributed to potential differences in the source or path effects that determine the waveforms. Nevertheless, catastrophe risk models consider only one of these events (in this case the first while the second is removed) for the computation of seismicity rates. This means that to replicate the damage potential of this sequence the stochastic catalog of these models will include only one event whose ground motion field covers a smaller area than that actually affected by the sequence. In other words, in the Emilia-Romagna example the stochastic catalog may include a simulated event

[1] http://shakemap.rm.ingv.it/.

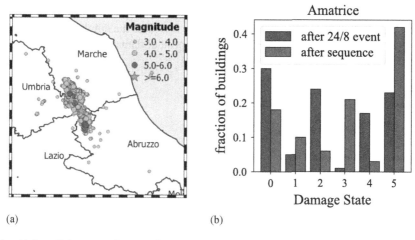

(a)   (b)

Fig. 11.2 a Epicenters of events belonging to the 2016–2017 Central Italy sequence, and **b** percentage of buildings in the town of Amatrice in different damage states immediately after the 24th August earthquake and after the entire sequence (Stewart et al. 2017a)

whose footprint of the ground motion used for damage and loss computation will be similar either to Fig. 11.1a or b but never to Fig. 11.1c. Hence, this approach will inevitably fail to take into account the contribution of foreshocks/aftershocks in the estimated maximum ground motion experienced at each site.

Another example that stands out is the 2016–2017 Central Italy sequence, which comprised thousands of earthquakes (11.2a), including nine M5 + earthquakes. Yet, if one were to apply a standard declustering procedure, such as the one proposed by Gardner and Knopoff (1974), the only significant event retained would be the October 30 M6.5 Norcia earthquake. In other words, as far as traditional seismic risk assessment is concerned, the August 24 M6.0 Amatrice earthquake, the October 26 M5.9 Visso earthquake and the six additional events with moment magnitude above 5 never happened, even though the damage they caused is well documented in Sextos et al. (2018) and also shown here in Fig. 11.2b. Another striking example outside of Italy is the 2010–2011 Canterbury earthquake sequence, which was started by the M7.1 Darfield earthquake followed by three large aftershocks with magnitude greater than 6 that devastated Christchurch more than the mainshock did.

In summary, from the viewpoint of single-site seismic hazard (and risk) assessment, what was stated above means that even though one can arguably claim that the rate of clusters is correctly estimated by counting mainshocks, the probability of ground motion exceedance given a mainshock rupture is most likely underestimated; its assessment does not factor in the impact of non-mainshock earthquakes unless, to a certain extent, a fault-specific geologic/geodetic approach based on seismic moment balance is followed. This results in seismic hazard (and risk) estimates that are biased low. This conclusion obviously carries to portfolio risk assessment. Therein, the spatial distribution of the exposed assets interestingly implies also that if a sequence involves multiple strong earthquakes, it is very likely that different assets at different sites are primarily affected by different earthquakes within the same cluster. Including all events in a sequence increases the number of potentially affected structures and enlarges the spatial footprint of damage.

11.2.2 Damage Accumulation

A different, more intuitive and perhaps more widely understood limitation of the current mainshock-only state-of-practice pertains to the so-called damage accumulation phenomenon. It is generally expected that structures in pre-existing damaged conditions owing to previous earthquakes are more prone to further deterioration, even if the following ground motions are weaker than the previously experienced ones. As a result, sites that experience more than one significant ground motion, i.e. sites within or in the vicinity of the epicentral areas of multiple events of a given sequence, are likely to experience losses that are higher than the losses that would be inflicted by the ground shaking of any of these events individually. The afore-mentioned Fig. 11.2b shows the percentage of buildings in the town of Amatrice that were assessed to be in damage states ranging from 0 (no damage) to 5 (collapse) during inspections just after the first significant 24th August event, as well as at the end of the entire 2016–2017 sequence, as reported by GEER (Stewart et al. 2017a). Even though the strongest ground motion in Amatrice was recorded during the initial 24th August shock, it is evident that the cumulative loading experienced during the sequence aggravated the condition of the building stock and even pushed a significant amount of already damaged buildings to collapse.

11.2.3 Arbitrariness in Declustering and Its Unintended Consequences

Aside from such conceptual defects of the mainshock-only view of seismicity for risk assessment purposes, there are additional motives to seek solutions to move past it. For instance, the process of declustering is highly subjective. The choice of the declustering algorithm that is incumbent on the analyst can lead to more or fewer earthquakes flagged as mainshocks, which in turn can result in strikingly different seismicity rates (Stiphout et al. 2011). Even within the boundaries of the same declustering technique, different analysts may choose different values of the parameters to delimit the spatio-temporal window for sifting earthquakes, which would lead to different numbers of retained earthquakes and, therefore, to different estimates of mainshock seismicity rates. Moreover, a declustering scheme that keeps the largest magnitude event of each cluster may lead to an unintended distortion of the magnitude-frequency distribution (Marzocchi and Taroni 2014). The latter is generally thought as exponential (Gutenberg-Richter law) for non-declustered catalogues, yet this shape is not fully preserved after declustering. The reason for this potential distortion is that largest events are likely maintained (as mainshocks) while lots of smaller magnitude events are filtered out. If an exponential function is then fitted to this modified catalogue, an underestimation of the b-value may take place. Smaller b-values in general lead to overestimation of hazard.

An additional concern regarding declustering is whether it really achieves its main purpose, i.e. to render the catalogue Poissonian (Luen and Stark 2011; Stiphout et al. 2011). Lastly, modeling seismicity as a time-independent mainshock-only process bears important practical limitations too. A particularly critical one is the inability to produce credible seismic hazard and risk estimates in periods of elevated seismic activity, such as during ongoing sequences or swarms of seismicity. Conventional time-independent, mainshock-only models are unable to pick up the increased likelihood of a new large earthquake and would yield unaltered loss predictions. This time independence of earthquake occurrence is acceptable, and also somewhat desirable, when the resulting seismic hazard is applied for underpinning design provisions in building codes that remain in force often for a decade or more and cannot change any time an earthquake occurs during the time they are adopted. But it is limiting in applications aiming at insurance and reinsurance pricing, where one is strictly interested in assessing risk for a short to mid-term time frame. Given that productive seismic sequences following large mainshocks may last for many years, this can render the results of seismic risk models deficient, if not unusable, for a significant amount of time.

11.2.4 Including Earthquake Sequences in Seismic Risk Assessment

Motivated by the above, there has been an increasing effort in recent years to commence quantifying the potential impact that foreshock, aftershock and triggered events have on seismic risk estimates. Some authors have attempted to assess the long-term seismic risk of individual buildings accounting for aftershock sequences (Jalayer and Ebrahimian 2016; Shokrabadi and Burton 2017), while a few have even tried to investigate the impact of seismicity clustering on short- and long-term portfolio risk assessment (Field et al. 2017; Papadopoulos and Bazzurro 2021; Shokrabadi and Burton 2019; Zhang et al. 2018). Figure 11.3 shows exceedance probability loss curves from the study of Papadopoulos and Bazzurro (2021), derived for the building stock of Umbria in Central Italy using a standard Poisson model, as well as representing seismicity via the Epidemic-Type Aftershock Sequence (ETAS) model (Ogata 1998). Given that the latter is a time-dependent model, its earthquake rates vary significantly depending on the past seismicity that is used as initial condition. The curves presented in the left panel of Fig. 11.3 refer to a one-year "active" period starting from 26/04/2017, i.e. about 8 months after the onset of the 2016–2017 Central Italy sequence, but only about three months after the last four M > 5 events. Thus,

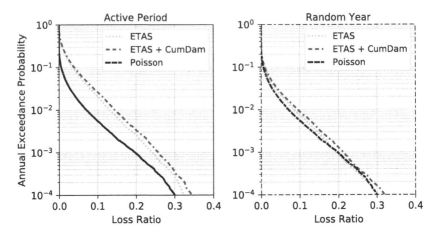

Fig. 11.3 Comparison of annual exceedance probability loss curves for the residential building stock in Umbria, obtained using a Poisson recurrence model and an ETAS model (adapted from Papadopoulos and Bazzurro (2021)). The left panel shows results referring to a specific active period following the 2016–2017 Central Italy sequence as described in the text, while the right panel refers to a random year of seismicity

it constitutes a period of increased seismic activity, which translates to significantly amplified risk estimates. On the other hand, the curves on the right panel refer to a random year of seismicity; therein the amplification of losses (due to the consideration of sequences) is still important, but clearly not as pronounced as in the former case.

Moreover, a first attempt to capture the effects of damage accumulation was also made. Damage-dependent fragility models, i.e. fragility functions referring to buildings with initial damage, were developed for a set of building classes by means of nonlinear single-degree-of-freedom (SDOF) system representations. For every rupture in the 20,000 realizations of the 1 yr-long stochastic catalogues used in the analyses, after the first event the damage states of the assets in the exposure dataset were sampled, stored and used as initial conditions for the damage estimation for the next event in the catalogue, if any. Given that each catalog covers only 1 year of seismicity, the assumption that no repair actually occurs in between earthquakes is certainly tenable. An alternative simplified computational workflow was also explored in which the same fragility curves were used regardless of the damage and loss sampled in previous events (but care was taken to avoid double counting losses by keeping track of the loss ratio of each building asset from previous earthquakes during the same year, if any). The loss curves obtained by explicitly accounting for damage accumulation and using the aforementioned simplified approach are shown in Fig. 11.3 in blue and cyan respectively, and compared against loss curves derived following a mainshock-only Poisson process based model. For more details on the loss estimation framework, the reader is referred to Papadopoulos and Bazzurro (2021). As expected, when accounting for damage accumulation the computed losses were found to be higher (Fig. 11.3), but its impact is smaller than that due to the consideration of earthquake clusters in risk assessment. This limited difference is due to the spatially limited effects of damage accumulation, which are significant only in the epicentral area of the largest event where many already severely damaged buildings are left to cope with additional shaking from other events in the sequence. In areas farther from the epicenter, the average damage level is low and the vulnerability to ground motion caused by successive earthquakes is essentially unchanged. Note that these findings cannot be generalized without further research and that the simplistic modeling of the complex phenomenon of damage accumulation may be somewhat coloring the impact of the phenomenon. However, the trend is clear.

In any case, while there is still no wide consensus on the methods and practices to utilize for modeling the spatio-temporal clustering of seismicity, most of the early studies suggest that aftershock sequences can have a moderate to large effect on seismic risk estimates. These early findings should arguably prompt researchers and catastrophe risk modelers to interpret and use the results of traditional risk models with caution. On the other hand, they also highlight the need for continued research to improve our capacity to more accurately capture the effects of seismicity clustering and begin refining earthquake risk models accordingly.

11.3 Why Identical Buildings at Different Locations have Different Vulnerability?

Vulnerability functions play a central role in regional seismic loss assessment for portfolios of structures (Calvi et al. 2006; Rossetto and Elnashai 2003) and are the main result of the vulnerability module of earthquake risk models. Based on the desired accuracy as well as the availability of the required data, four different approaches have been utilized for the development of such functions for portfolio loss estimation: (1) expert judgment, (2) empirical, (3) analytical/mechanical, and, (4) hybrid (e.g., empirical plus analytical) (Kappos et al. 1995; Kappos et al. 1998; Barbat et al. 1996; Akkar et al. 2005; Bommer and Crowley 2006) methods. The *first approach*, widely used in the early days (ATC-13 1985; Brzev et al. 2013), has been abandoned, in the sense that it is no longer used in isolation but only to validate the reasonability of the results derived using other approaches. The *second approach* uses damage data collected in damage reconnaissance missions after an earthquake and loss data from insurance claims to generate vulnerability functions (Orsini 1999; Rossetto and Elnashai 2003; Di Pasquale et al. 2005; Porter et al. 2007; Straub and Der Kiureghian 2008; Rossetto et al. 2014; Noh et al. 2015). This is usually the preferred route provided that enough usable damage/loss data are available for a specific class of structures (e.g., midrise reinforced concrete frames of the 1970s) and that the ground motion experienced at the sites where the damaged structures are located can be estimated with a reasonable accuracy. This approach, however, is seldom practicable and even when it is, it is only applied to some specific building classes (e.g., wood frame buildings in California) because post-event specific data for each building class are never collected in such a clean fashion to make it applicable across the board. To the authors' knowledge the second approach has never been applied alone to all construction classes in any earthquake risk assessment model without resorting to some support from analytical studies. In the *third approach*, numerical analyses (Kennedy and Ravindra 1984; Porter et al. 2014; Silva et al. 2015; D'Ayala et al. 2014) carried over on computerized models of representative archetype (FEMA-P695 2009) or index buildings in each class provide simulated losses to replace, or supplement, missing or scarce empirical data. Of course, there are numerous cases, for example with newer buildings that never experienced a damaging earthquake, where clearly the analytical approach is the best, if not the only, viable option.

11.3.1 Vulnerability Functions based on the Analytical Method

In applying the analytical method, which is by far the most widely adopted in practice, for each class of buildings (or other types of structures, such as bridges) engineers generate finite element models of one or more structures using either a nonlinear single-degree-of-freedom (SDOF) or a multi-degree-of-freedom (MDOF) representation. Consequently, based on the required level of accuracy, structural analysis and vulnerability analysis methods are employed to assess their response to different levels of ground motion. The utilized structural analysis approaches are either nonlinear static analysis such as capacity-spectrum (FEMA 2003; Sousa et al. 2004; Calvi and Pinho 2004) plus displacement-based methods (Pinho et al. 2002; Restrepo-Vélez and Magenes 2004) or nonlinear dynamic analysis (Haselton et al. 2011; Jayaram et al. 2012; Silva et al. 2014), which typically uses ground motion recordings from past earthquakes, as discussed later. In modern, last-generation models, however, the simpler but more approximate nonlinear static analysis has been largely abandoned in favor of its more accurate dynamic counterpart. Hence, in the following we only focus on the issues related to using the nonlinear dynamic analysis for developing analytical vulnerability functions.

Vulnerability analysis approaches use a single vulnerability function for assessing the performance of the entire building. The building performance is measured either by means of *global response metrics*, or by looking at the performance of each *specific building component*. In the latter case, the response of a component depends on its location and, therefore, it is gauged by *story-specific response metrics*, (Porter et al. 2001; Mitrani-Reiser 2007). The component-based methodology (FEMA-P58 2012), whose applications can be found in several studies (Porter et al. 2001; Mitrani-Reiser 2007; Kohrangi et al. 2016b), is arguably superior but it is more time consuming because it requires detailed information about the location, the damageability and the repair cost of all the structural and non-structural components and contents of the building. These methodologies were originally devised for a single building located at a specific site. However, they are also applied with little or no modifications in portfolio loss estimation to one or more archetypes representing an entire class of buildings that could be located anywhere within a region. This is an important observation that is central to the discussion below.

11.3.2 Vulnerability Functions for Single Buildings and for Building Portfolios: The Present

In practice, for portfolio loss estimation, for any building class in a given country an engineer either.

(a) selects and adopts vulnerability functions available in the literature (perhaps based on data not from the same region of interest) often without an in-depth scrutiny, or

(b) develops new numerical functions (FEMA 2003) using the global or the component-based approach but without paying enough attention to the ground motions used to estimate the response.

The *former route* should be followed with extreme caution. Even if two buildings in two different countries could be confidently categorized into the same class (e.g., low-rise ductile RC frame building), which is not always the case, their vulnerability functions would likely differ, sometimes considerably. Buildings in different parts of the world are, in fact, the final results of different design codes and construction practices. Hence, they naturally have different structural characteristics and seismic performance. Therefore, analysts usually develop region or country-specific vulnerability functions (Bal et al. 2008; Villar-Vega et al. 2017). Note that even in those cases where building codes are similar in two countries, the performance of two like buildings may still not be similar due to the different levels of code enforcement and construction quality assurance that may exist in the two countries. The custom of utilizing vulnerability functions from other countries is clearly a potential source of bias on the resulting loss estimates.

The *latter route*, which is clearly preferable, is not free of hurdles either. For many years, as alluded to earlier, it has been a common practice in the regional loss assessment to select one (or more) archetype building in a class and to assess its response using a set of ground motions. The set of recordings, which is typically selected without particular care to the local seismicity and level of hazard in the country, is utilized as input to perform Incremental Dynamic Analysis, IDA (Vamvatsikos and Cornell 2002) or some form of multi-stripe or cloud analysis (Jalayer 2003). The resulting vulnerability function is then applied to all the buildings in the class regardless of where they are located in the country.

However, the application of this methodology, originally intended for a given building at a given site, has shown that reliability of the resulting vulnerability function is achieved only when ground motions consistent with the hazard at the site are selected and when the ground motion IM utilized to predict the vulnerability is carefully chosen. This is a generally less understood but very important aspect. In simple words, the vulnerability function of a building does not only depend on the building itself but also on the characteristics of the ground motions that the building may experience in its lifetime. The reason for this dependence is hidden in the way vulnerability functions are constructed. Vulnerability functions are based on only one ground motion IM (almost always the first-mode spectral acceleration, SAT_1, or the peak ground acceleration, PGA) leaving everything else unaccounted for. Given the same level of IM, however, the other characteristics of the ground motions, such as their spectral content and duration that do affect the response, are systematically different at different sites. They differ because they depend on the parameters (e.g., magnitude and distance of the building site from the main seismogenic sources) of the local earthquakes that most contribute to the hazard at a given site. In general, different sites have different controlling earthquakes that generate ground motions with different unaccounted-for characteristics that, in turn, cause vulnerability functions of identical buildings at different sites to be distinct.

Figure 11.4 shows the profiles of two commonly used building response measures for vulnerability assessment, the median inter-story drift ratio (IDR) and peak floor acceleration (PFA), of an identical 7-story building located in three different cities in Turkey: Istanbul, Ankara and Erzincan. The profiles shown for these three cities are the mean values obtained by running three different sets of ground motions with the same intensity of $SAT_1 = 0.35$ g (Kohrangi et al. 2017b), but with spectral shapes consistent with the hazard of each one of the three sites. The hazard consistency was imposed via the Conditional Spectrum (CS) method (Jayaram et al. 2011; Kohrangi et al. 2017a), which along with the Generalized Conditional Intensity Measure, GCIM, (Bradley 2010) method form the state-of-the-art for performing site-specific hazard-consistent record selection. The profiles for both IDR and PFA

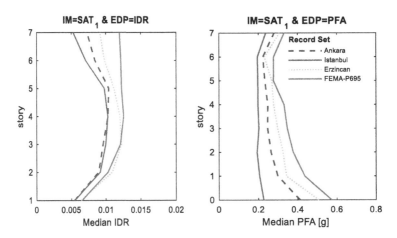

Fig. 11.4 Inter-story drift ratio (IDR) and peak floor acceleration (PFA) profiles of an identical 7-story RC building located in three different cities in Turkey. Legend: EDP = Engineering Demand Parameter

at the three sites are clearly different, a difference that is lost in all regional portfolio loss assessment studies performed nowadays. As explained earlier, in practice, to develop vulnerability functions for archetype buildings, an engineer would take a readily available set of ground motions, such as the FEMA-P695 records, and carry on response analysis. The results of this exercise for the same 7-story building subject to the far-field record set of FEMA-P695 scaled to $SAT_1 = 0.35$ g are also shown in Fig. 11.4. In this particular case, by chance, the FEMA-P695 records (all scaled to the same $SAT_1 = 0.35$ g as done for the other three city-specific record sets) appear to generate more severe demands in the structure for both IDR and especially PFA than the other three sets. The larger damageability of this FEMA-P695 set holds also for IDR and PFA profiles computed for records scaled to other SAT_1 levels. If this FEMA-P695 set of records were used to derive a vulnerability function for this archetype 7-story RC building and this function were used to estimate losses for all midrise RC buildings of that era in Turkey, the losses would be severely over-estimated for all such buildings especially those in the two most heavily populated cities of Istanbul and Ankara. Any other random set of ground motions other than the FEMA-P695 would have generated different vulnerability functions with unknown reliability.

This simple exercise clearly supports the necessity of using ground motions that are site-hazard consistent and the consistency can be enforced for specific sites using CS, as we did, or GCIM. In a portfolio loss assessment, however, there are thousands of sites and developing one vulnerability function per site for each class of buildings would be highly impractical. The alternative question here is: "How can we make a ground motion record selection that is consistent with the seismic hazard in a region and not at a single site?" This selection would lead to a vulnerability function that is, on average, appropriate for all buildings in the region of interest.

11.3.3 Vulnerability Functions for Building Portfolios: The Future

The best practical strategy to tackle this problem is to develop vulnerability functions based on a ground motion IM that has the highest possible predictive power of the structural response engineering demand parameters (EDPs) that are used as building response metrics. This ideal IM would also need to be "predictable" given the usual parameters of the causative earthquake, such as magnitude and rupture mechanism, and of the site, such as local soil conditions and distance from the rupture. If this ideal IM were identified, an engineer would be able to predict with very little uncertainty the values of all the EDPs in a building caused by a ground motion with only the knowledge of the value of such an ideal IM. In the literature this ideal IM would be called perfectly "sufficient" (Luco and Cornell 2007). In this hypothetical case, ground motions from different regions with different durations and different causative magnitude earthquakes but with the same value of this ideal IM would generate the same distribution of building response. In simple terms, this ideal IM would make consistency with the site hazard in the ground motion record selection essentially irrelevant for building response assessment. In the context of portfolio analysis, the identification of this IM would remove the dependency of the vulnerability function from the selection of the site-specific ground motions, which is evident in Fig. 11.4 when SAT_1 was used as the conditioning IM.

In the last decade, many studies (Cordova et al. 2000; Tothong and Luco 2007; Bianchini et al. 2010; Eads et al. 2015; Eads et al. 2013; Kazantzi et al. 2015; Vamvatsikos and Cornell 2005; Kohrangi et al. 2016a) focused on application of advanced IMs to reduce, to the extent possible, the dependency of the response and, therefore, of the vulnerability function on the site hazard. In particular, after the promising results in the assessment of site-specific building-specific losses (Kohrangi et al. 2017a), the authors proposed a multi-site record selection scheme for developing vulnerability functions for portfolio loss assessment that is both practical and reasonably accurate (Kohrangi et al. 2017b). This approach uses the average spectral acceleration, $AvgSA$, as an advanced IM for response prediction while a modification of the CS method to account for $AvgSA$ at multiple sites is used to ensure the hazard consistency at the regional, rather than at the site, level. The regional hazard consistency is obtained by the law of total variance, which incorporates the impact of multiple sites into a single record set. The weight associated to each site, for example, was chosen to be proportional to the replacement cost of that building class at that site as a fraction of the total replacement cost of that class in the entire country. This method provides a unique set of ground motion records that has a balanced contribution from the ground motions relevant to the hazard at the different sites to use as input to the vulnerability analyses computations. Note that the amount of response analyses that the engineer needs to carry out is identical to before, only the records are selected more judiciously.

Figure 11.5b and d show the results of this $AvgSA$-based multi-site-hazard-consistent approach for our illustrative example entailing the same 7-story reinforced concrete building located at the three sites of Istanbul, Ankara and Erzincan. More specifically, Fig. 11.5b shows a comparison of collapse fragility curves and Fig. 11.5d displays the vulnerability functions obtained using ground motions records selected using the site-specific approach for the three cities, the proposed multi-site $AvgSA$-based approach discussed above, and the FEMA-P695 set (this last one scaled as done in an IDA framework to multiple IM levels). The multi-site fragility and vulnerability functions are, for all practical purposes, almost indistinguishable from those developed specifically for the three cities. This means that utilizing this approach to develop fragility and vulnerability functions can potentially remove the existing bias from portfolio loss estimates. It is also apparent that the use of a much more sufficient IM, such as $AvgSA$, makes the family of fragility curves to be very tight regardless of the method used to select the input ground motions. Figure 11.5a and c show the fragility and vulnerability functions, respectively, computed using the same multi-site-hazard-consistent methodology of Kohrangi et al. (2017b) but this time based on the poorly "sufficient" SAT_1, as done routinely in practice, rather than on $AvgSA$. The families of curves are much more dispersed and the proposed multi-site

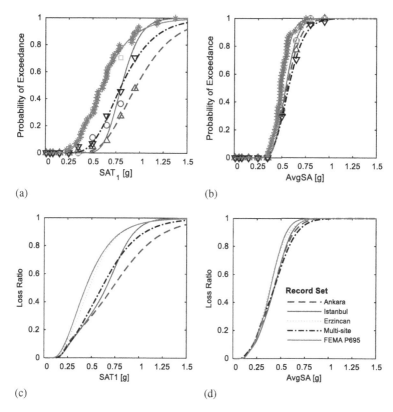

Fig. 11.5 Collapse fragility curves (top panels) and vulnerability functions (bottom panels) of a 7-story reinforced concrete frame building located at three sites in Turkey using two different conditioning IMs: SAT_1 in (**a**) and (**c**); $AvgSA$ in (**b**) and (**d**). Adopted from Kohrangi et al. (2017b)

approach is less efficient in reducing the bias that may stem from using randomly or, in any case, poorly selected ground motion records.

The effects on the risk estimates of the IM chosen for developing vulnerability functions and of the record selection scheme selected for the computations of the vulnerability functions are shown in Fig. 11.6. This figure shows the different estimates of the mean annual rate of exceedance loss curves for the 7-story reinforced concrete building discussed so far located in the three cities of Ankara, Erzincan, and Istanbul. These curves were computed by convolving the hazard curves for soft soil conditions for these three cities with the corresponding vulnerability curves in Fig. 11.5c and d. Different important observations stem from the inspection of Fig. 11.6. The first one is certainly the difference in the site-specific loss exceedance curves for the same city obtained using SAT_1 and $AvgSA$ as the pivotal IM. The difference stems from the lack of sufficiency especially of SAT_1 and, to a lesser extent, of $AvgSA$ but, as argued in Kohrangi et al. (2017a), the $AvgSA$-based curves are statistically more robust and closer to the true but unknown curve. Therefore,

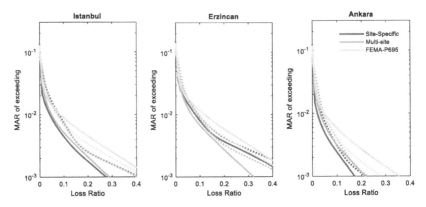

Fig. 11.6 Mean annual rate of exceedance loss curves obtained for the same 7-story reinforced concrete building located in Istanbul, Erzincan and Ankara using different approaches for the ground motions selected for the development of the vulnerability function. Legend: solid line: SAT_1; Dotted line: $AvgSA$

for the discussion at hand, the benchmark curve for the building in each city should be considered the site-specific curve based on $AvgSA$. The second observation is on the variability between the loss exceedance curves obtained for the same city using the different approaches. This variability is much larger for the family of curves based on SAT_1 than for that based on $AvgSA$. Selecting a set of records for vulnerability computations that is not consistent with the site and regional hazard, such as the FEMA-P696 set for example, can lead to severe bias in the risk estimates. In the case at hand, by chance that record set would have provided reasonably accurate results only for Erzincan but severe overestimation of the risk for the much more important cities of Istanbul and Ankara. All these differences, however, almost vanish when $AvgSA$ is used for the development of vulnerability functions. Because of the much higher sufficiency of this IM, even the selection of the records used for developing vulnerability curves becomes much less important. The estimates of the loss exceedance curves are closely clustered around the benchmark regardless of the ground motion records selection scheme adopted.

11.3.4   Final Remarks

To conclude, the current practice of developing vulnerability functions via the analytical method for classes of buildings is the most widely used in commercial models for portfolio loss estimation. This practice, which hinges on using inferior and insufficient IMs, such as SAT_1, to predict structural responses coupled with a primitive ground motion record selection is prone to introducing bias of unknown amount and sign in the final loss estimates. The use of $AvgSA$, which is a significantly more sufficient IM, coupled with a careful record selection that balances the characteristics of ground motions at the sites of interest in the study region is surfacing as a much more promising approach for removing such a bias. As an intermediate step, with limited effort the current practice could embrace the use of the SAT_1-based multi-site-hazard-consistent approach, which by virtue of producing vulnerability functions that are balanced in terms of ground motion characteristics likely expected at the different sites of the study region, may at least reduce, if not eliminate, the bias in the loss estimates.

11.4 Beyond Ergodic Seismic Hazard Estimates and Impact on Risk

A core component of probabilistic seismic hazard is the ground-motion prediction equation (GMPE). For any prospective earthquake, in the second part of the hazard module of any earthquake risk assessment model a GMPE is required to predict the distribution of any (log) ground-motion intensity measure, IM, at any site of interest in the affected region. The (log) IM at a given site is modeled as a Gaussian distribution $N(\mu, \sigma)$, where μ is the median value of the IM and σ is the standard deviation given the parameters of the earthquake, typically the earthquake moment magnitude (M_w), the distance of the rupture from the site (e.g., the Joyner-Boore metric R_{JB}), and the site conditions (often described by the time averaged shear-wave velocity Vs_{30} in the top 30 m). However, to obtain empirical GMPE that yield reliable and stable IM median predictions even for scenarios beyond those ever recorded at a site, the developers apply statistical techniques on datasets of ground-motion observations from a variety of earthquakes recorded at a multitude of sites scattered across the globe. This practice, which is dictated solely by the scarcity of data at any single site, essentially substitutes long-term sampling of ground motions at a given site with short-term sampling at several other sites with similar conditions. In statistical parlance, this is the so-called ergodic assumption in GMPEs (Anderson and Brune 1999). The ergodic assumption predicates that the *generic* μ of a GMPE predicts theoretically and physically constrained ground motions over a wide range of magnitude, distance, and site conditions. In other words, ground motions from earthquakes occurred somewhere else in the world and recorded at sites similar to the one considered are acceptable proxies for the ground motions that local earthquakes could cause at the specific site of interest. In this ergodic framework, all the spatiotemporal *variabilities* of the geophysical properties unaccounted in the ground-motion prediction are considered *natural randomness*, and are relegated to the aleatory variability σ of the IM. These geophysical elements of a seismic process can be broadly identified as those related to the physics of earthquake ruptures, shear-wave propagation paths, and receiving sites' conditions.

Recent years have witnessed an exponential growth of ground-motion observations from several seismically active regions of the world. GMPEs have benefitted tremendously from these high-quality datasets, especially in reducing the epistemic (modelling) uncertainty of μ at a given site for a wide variety of scenarios. However, in a review of the 50 years of GMPE development, Douglas (2014) reported a counter-intuitive increase in σ, namely an increase in the unresolved spatiotemporal variability of seismic processes. In a probabilistic seismic hazard assessment framework, an increasing σ implies an increasing likelihood of rare (and large) ground motions from even moderate-sized earthquakes (Bommer and Abrahamson 2006). In turn, this implies the paradox that, despite an increasing amount of high-quality ground-motion data, the ergodic hazard estimates become more uncertain and, in turn, the associated likelihood of observing large damage and losses due to earthquakes increases. To mitigate the negative consequences of this issue, GMPE developers have started opting out of the ergodic assumption to focus on explaining some of the apparent natural randomness of (geophysics of) seismic processes. In other words, the shift towards non-ergodic GMPEs has started.

The ergodic assumption can be relaxed with the discovery of repeatable seismic phenomena in the data and in explicitly modelling them into the μ of the GMPE, as discussed in the next section. This operation changes the median estimate from the value of the ergodic one and, at the same time, reduces the aleatory variability σ. Figure 11.7 is an illustration of shifting from a generic ergodic prediction $N(\mu, \sigma)$ to three non-ergodic level i specific predictions with their own unique distributions $N(\mu_i, \sigma_i)$; where levels $i = 1, 2, 3$ could be ruptures, paths, sites, or any spatiotemporal-specific (characteristic). Plainly put, the more non-ergodic a GMPE becomes, the more spatio-temporally specific its μ_i, and lower its σ_i will become.

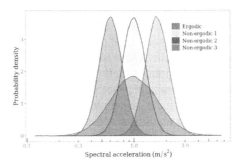

Fig. 11.7 Illustrative example of the distribution of an IM from a scenario event at four sites with identical Vs_{30} and distance from the causative earthquake. The effects on the IM distribution in the three non-ergodic cases due to the reduction in σ and shift in μ is evident. Courtesy: Dr. Norman Abrahamson

11.4.1 Partially Non-ergodic GMPEs

In the quest for non-ergodic GMPEs, it is first necessary to identify the three distinct pieces that contribute to shaping the ground motion features recorded at any given site, namely.

1. earthquake ruptures and their seismogenic sources,
2. shear-wave propagation paths and their regions, and
3. receiving sites' geology and topography.

For appropriately modeling them, the second step involves the characterization of their spatiotemporal specifics, as discussed below.

Although the interaction between these components is complex, the pursuit for non-ergodic GMPEs has been quite rewarding in the past few years. Anticipating fully non-ergodic GMPEs, Al Atik et al. (2010) proposed a taxonomy for the components of σ (McGuire 2004), which was complemented by several follow-up studies leading up to the most recent one by Baltay et al. (2017).

11.4.1.1 Characterization of Earthquake Ruptures and Their Seismogenic Sources

Among the various parameters characterizing an earthquake rupture, most GMPEs use only the M_w as a predictor variable in the functional form for estimating μ of the ground-motion IM at a site. However, M_w is only indicative of the size of an earthquake rupture, and not as much of the amount of shear-wave energy radiated from the elastic rebound, which is the actual cause of ground-motion at a site. This radiated energy is best correlated to the tectonic stress released by a single rupture, and is (generally) quantified as stress-drop in the rupture's Brune (1970) shear-wave spectrum.

Depending on their spatial and temporal origins, earthquake ruptures of identical M_w may show very dissimilar stress-drops, e.g., see Cotton et al. (2013). This may arise from the differences in the tectonic regimes, crustal deformation rates, periodicity of stress release, and other tectonic processes that lead to a rupture. For example, a recent study by Bindi and Kotha (2020) showed that the M6.5 Friuli earthquake (1976) had substantially higher stress-drop than the recent M6.5 Norcia earthquake (2016), which in-turn, released higher energy than the more recent and nearby M6.3 L'Aquila earthquake (2009). Therefore, the functional form for predicting μ of a non-ergodic *rupture-specific* GMPE would use at least the rupture-specific stress-drop as a predictor variable. This enhancement leads to a substantially lower value of the non-ergodic σ (Bindi et al. 2018b, 2019).

In practice, however, the stress-drop of even the largest ruptures is hard to predict. Therefore, a reasonable level of non-ergodic *source-specificity* can be achieved by spatially localizing several ruptures and their stress-drops to seismogenic sources;

Fig. 11.8 Map illustrating the spatial variability of locality-specific ground-motion prediction adjustments to the ergodic median IM, here PGA. Red polygons are tectonic localities generating earthquake ruptures that produce a median PGA higher than the ergodic median, while blue polygons correspond to those producing a median PGA lower than the ergodic median, as estimated from the Engineering Strong Motion dataset by Kotha et al. (2020)

such as fault systems, hypocentral depths (Abrahamson et al. 2014), tectonic localities (Kotha et al. 2020), epicentral coordinates (Landwehr et al. 2016), or epicentral neighborhood (Lin et al. 2011). For example, Fig. 11.8 is a result from the recently developed partially non-ergodic GMPE of Kotha et al. (2020) from the Engineering Strong Motion dataset (Lanzano et al. 2018; Bindi et al. 2018a). In their study, Kotha et al. (2020) grouped the 927 shallow crustal ruptures of $3.0 < M_w \leq 7.4$ occurred between 1976–2016 in the pan-European region into tectonic localities (groups of seismic sources) devised in the purview of European Seismic Hazard Model 2020 (ESHM20). These localities are larger regions containing tectonically similar seismic sources. In developing the locality-specific GMPE, the systematic differences between the ground-motion IMs caused by earthquakes occurred in various localities are quantified into a quantity called $\delta L2L_l$, where l indexes one of the locality polygons shown in Fig. 11.8. Ruptures originated in the red localities have produced, on average, stronger ground motions than those occurred in the blue localities. With respect to the ergodic median of the GMPE, the median IM of the former ground motions was higher than the ergodic median (like Non-Ergodic Case 2 in Fig. 11.7) and vice versa for the latter (like Non-Ergodic Case 3 in Fig. 11.7). Such estimates, if assumed temporally stationary as customarily done, can be used to predict partially non-ergodic locality-specific ground-motion IMs for prospective ruptures.

Geologists and geophysicists have recently invested considerable energy in identifying parameters that can help characterize the spatiotemporal variability of seismogenic source properties (see the localities in Fig. 11.8), such as fault maturity (Radiguet et al. 2009), seismic moment-rate density (Weatherill et al. 2016), and rupture velocity (Chounet et al. 2018). On the other hand, the temporal variability of rupture characteristics has proven more difficult to resolve empirically, because it can be associated to the characteristics of the propagation medium (e.g., crustal velocity structure), or to the characteristics of the source properties (e.g., fault healing), or to some combination of both. Nevertheless, assimilation of large amounts of ground-motion data from active seismic sources helps resolving at least the spatial variability, while assuming temporal stationarity for now.

11.4.1.2 Characterization of the Shear-Wave Propagation Paths and Their Regions

While source-specific, in lieu of rupture-specific, predictions are still in development, the investigation of the spatial variability of *shear-wave propagation effects* has seen a much earlier start, e.g. Douglas (2004). Ergodic GMPEs typically predict ground-motions that decay identically with distance towards the site regardless of the region of the world where they are applied. Several GMPEs have, however, already quantified the regional differences in attenuation of ground-motion with distance. The lithospheric properties, such as shear-wave velocity and depth to Moho, vary rapidly across the seismically active regions. To account for these variations, for instance, Kotha et al. (2016) proposed a partially non-ergodic *region-specific* GMPE for the pan-European regions, where a region encompasses several possible paths of shear-waves. The region-specific μ of that GMPE distinguishes the much faster decay (with distance) of high-frequency ground-motion IMs in Italy compared to that in the rest of Europe. Similarly, the GMPE by Kale et al. (2015) quantified regional differences in attenuation between Iran and Turkey; while Boore et al. (2014), Campbell and Bozorgnia (2014), and Abrahamson et al. (2014) models distinguish attenuation between California, Japan, New Zealand, China, and other active regions.

More recently, using the 18,222 records in the ESM dataset,[2] Kotha et al. (2020) quantified the regional differences in apparent anelastic attenuation of IMs across 46 regions of the TSUMAPS-NEAM regionalization model (Basili et al. 2019) spanning the most seismically active regions of pan-Europe. The systematic differences across regions in the anelastic attenuation are quantified in the quantity $\delta c_{3,r}$ mapped in Fig. 11.9. With respect to the median of high-frequency IM of the ergodic GMPE, shear-waves traversing the blue regions in Fig. 11.9 experience a faster anelastic decay (i.e., lower median), while those traversing the red regions propagate more efficiently (i.e., higher median).

11.4.1.3 Characterization of Receiving Sites' Geology and Topography

The momentous shift towards the development of non-ergodic GMPEs has been propelled by the need of resolving the third piece of the puzzle, namely the *site-to-site variability* of ground-motion characteristics. Anderson and Brune (1999)

---

[2]https://esm.mi.ingv.it//flatfile-2018/.

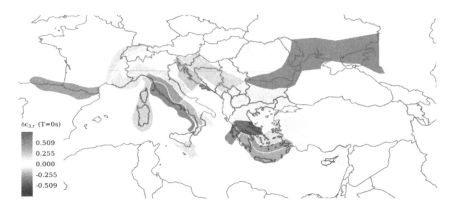

Fig. 11.9 Map illustrating the spatial variability of region-specific ground-motion adjustments to the ergodic median IM (here PGA). Red polygons are regions with weaker than average anelastic attenuation of PGA (i.e., higher median PGA), while blue polygons correspond to regions with anelastic attenuation of PGA stronger than global average (i.e., lower median PGA), as estimated from the Engineering Strong Motion dataset by Kotha et al. (2020)

triggered the need to resolve ergodicity of GMPEs by first identifying the epistemic and aleatory components of ground-motion variability. Assuming a specific site's response to seismic action is temporally invariable (i.e., site's local soil and topographic conditions do not change dramatically between two consecutive earthquakes), a sufficient number of recordings at a given site would allow quantifying the *site-specific* response with reasonably low epistemic uncertainty. This uncertainty would also decrease asymptotically with more recordings. At such sites with multiple recordings, the aleatory site-to-site response variability, of course, does not apply anymore and during ground motion prediction it can be removed from the ergodic σ of the GMPE. This reduced non-ergodic σ, sometimes called the single-site σ in the literature, can be 30 to 40% smaller (Rodriguez-Marek et al. 2013) than its ergodic counterpart.

In addition to the source (Fig. 11.9) and path (Fig. 11.8) specific adjustments to ergodic GMPEs, the increase in high quality ground motion data has also allowed developing site-specific adjustments for many locations with multiple recordings. For example, Kotha et al. (2020) compiled site-specific (Fig. 11.10) adjustments for 1,829 sites in the ESM dataset, out of which 1,047 have recorded more than 3 earthquakes. In this figure, the locations of sites with red color are those at which the median of the recorded PGA values was systematically higher than the generic median of the ergodic GMPE for the same scenario earthquakes. The opposite trend applies to the blue marked locations. The development of such non-ergodic site-specific GMPEs and their application in seismic hazard and risk assessment are elaborated in a few recent studies (Kotha et al. 2017; Faccioli et al. 2015; Rodriguez-Marek et al. 2013), where the quantified differences between ergodic and non-ergodic site-specific assessments are shown to be enormous.

Fig. 11.10 Map illustrating the spatial variability of site-specific ground-motion adjustments. Red markers are sites whose median of the recorded PGA values for past earthquakes was larger than the ergodic median PGA. The opposite holds for the median PGA at the blue marker locations. The adjustments, expressed in terms of the $\delta S2S_s$ quantity mapped here, were estimated from the Engineering Strong Motion dataset by Kotha et al. (2020)

11.4.2 Effects of Partially Non-ergodic GMPEs on Risk Estimates

Following the development of partially non-ergodic GMPEs, the seismic hazard assessment community has started adopting these into practical applications (Weatherill et al. 2020; Stewart et al. 2017b; Walling 2009). A first attempt in assessing the impact on risk estimates due to shifting from ergodic to more accurate partially non-ergodic GMPEs adjusted for site-specific ground-motion predictions was done by Kohrangi et al. (2020) for single structures located at three sites in Turkey. This study considered elastic-perfectly plastic SDOF systems with initial elastic periods of $T_1 = 0.2$ s, $T_1 = 0.5$ s and $T_1 = 1.0$ representing ductile moment-resisting-frame buildings with different low-to-medium heights. The three sites (Table 11.1) were chosen based on their Vs_{30} values reported in the RESORCE dataset (Akkar et al. 2014) to represent rock sites (Site #1), very stiff sites (Site #2), and stiff sites (Site #3) as classified in the Eurocode 8. Table 11.1 lists also the number of ground-motion

Table 11.1 Three selected sites from the RESORCE database. Courtesy: Kohrangi et al. (2020)

Site #	Station #	Vs_{30} [m/s]	Latitude [Degree]	Longitude [Degree]	N_{obs}[a]	PGA [g]
1	"4802"	747	37.033	27.440	8	0.02–0.22
2	"4401"	481	38.350	38.340	10	0.01–0.18
3	"777"	339	41.020	28.950	8	0.03–1.72

[a]The number of observations used for deriving the site-specific adjustments for PGA. This number may be lower than the values reported for computing the adjustments for long-period SAs due to the lower bound of the usable frequency ranges of some of the records

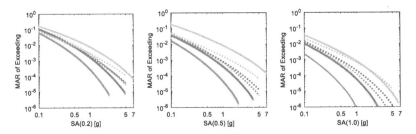

Fig. 11.11 Comparison of SA at $T_1 = 0.2$, 0.5, 1.0 s (left to right columns) ergodic (dotted) and non-ergodic site-specific (solid) hazard curves (top-row) and ductility exceedance risk curves (bottom-row) for identical buildings at the three sites (color coded). Courtesy: Kohrangi et al. (2020)

recordings used in estimating their site-specific adjustments, i.e. their site-specific $\delta S2S_s$ values as done for the sites in Fig. 11.10.

Kohrangi et al. (2020) compared the hazard and risk estimates at these three sites obtained using the area source seismicity model of SHARE (Woessner et al. 2015) and the two versions of the Kotha et al. (2016) GMPE. The ergodic version relies on the sites' Vs_{30} (a proxy of soil stiffness) to predict *soil-specific* (but site-generic) ground motions, while the non-ergodic version ignores Vs_{30} and uses the site-specific empirical amplification factors $\delta S2S_s$ to predict *site-specific* ground motions. The ergodic (dotted) and non-ergodic (solid) hazard curves at the three sites, are compared in Fig. 11.11. It is evident that the differences between ergodic and non-ergodic hazard estimates are site-dependent, period-dependent, and hazard-level-dependent. For instance, the site-specific hazard curves at rock Site #1 are significantly below the ergodic estimates, a trend that reflects the stronger local deamplification of short and long period ground motions at Site #1 compared to other sites (in the dataset) with similar Vs_{30}. Meanwhile, at the stiff soil Site #3, moderate period ($T_1 = 1.0$ s) ergodic and non-ergodic hazard curves are almost identical, but differ significantly at $T_1 = 0.5$ s.

The results of the response analyses carried out on these three structures using site-specific hazard consistent sets of ground motion records led to the vulnerability curves shown in Fig. 11.12. The risk estimates for these three structures at the three sites displayed in the form of loss exceedance curves in Fig. 11.13 were obtained via convolution of these vulnerability curves and the hazard curves shown in Fig. 11.11. The impact due to the more accurate site-specific non-ergodic GMPEs is evident. In general, the traditional ergodic GMPE would lead to a severe underestimation of the risk for all three structures at Site #3 but to a lesser extent for the more flexible one than for the stiffer one. Vice versa for Site #1 where the ergodic GMPE would lead to an overestimation of the risk. At Site #2 the ergodic GMPE yields very similar but slightly conservative risk estimates compared to the more precise estimates obtained via the site-specific non-ergodic GMPE.

Hence, as concluded by Kohrangi et al. (2020), this example shows that the traditional approach may lead to biased risk estimates whose amplitude and sign are impossible to predict a priori unless high quality site-specific ground-motion data

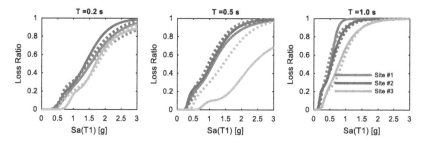

Fig. 11.12 Vulnerability functions for the three structures with fundamental period of vibration of $T_1 = 0.2$, 0.5, 1.0 s located at the three sites. Legend: solid line: hazard estimates based on site-specific non-ergodic GMPE; Dotted line: hazard estimates based on ergodic GMPE

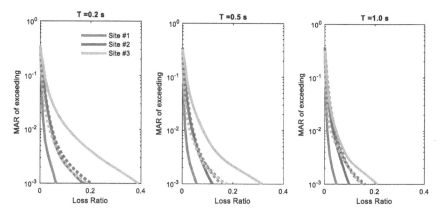

Fig. 11.13 Mean annual rate of exceedance loss curves obtained for the three structures located at the three sites. Legend: solid line: hazard estimates based on site-specific non-ergodic GMPE; Dotted line: hazard estimates based on ergodic GMPE

(Bard et al. 2019) allow the development of site-specific non-ergodic GMPEs. Based on these results, the use of the non-ergodic approach is recommended, whenever existing data allow it. However, further advancements of non-ergodic GMPEs are necessary before being routinely utilized in real life risk assessment applications.

11.5 Sources of Bias in Pricing of Earthquake Insurance Policies

As mentioned earlier, one of the key decisions that hinges on the results of catastrophe risk models is the determination by insurance and reinsurance companies of how much premium is fair to charge to cover the cost of an insurance product and generate sufficient profit. In insurance parlance this activity is called *pricing*. The price that a customer pays for the product, namely the market price, is a function of the so-called technical price, which can be either higher or lower of the market price depending on internal business strategies of the company. In turn, the technical price consists of:

1. The pure premium, which is the expected loss that the insurer can expect to pay, on average, every year over a long period of time. This quantity, is simply the model-based estimate of the Average Annual Loss (AAL) mentioned in the introduction.
2. Expense loading (included to account for internal operational costs, taxes, fees, commissions, reinsurance and retrocession costs, cost of capital, etc.)
3. Profit loading
4. Risk loading (to account for unmodeled perils and unknowns)

In this section we discuss the main contribution to the technical price, namely the pure premium, which is the only part of a technical price that has a scientific basis Figs. 11.14, 11.15 and 11.16 compare the AALs for the structures considered in the examples discussed in the three main sections of this chapter. These AALs were obtained using both the traditional approach and the enhanced approach suggested in those sections.

More specifically, Fig. 11.14 shows the AAL estimates for the active and random seismicity years computed for the building stock of Umbria using mainshock only seismicity (i.e., the traditional approach) and clustered seismicity that includes all other events. Not unexpectedly given the different loss exceedance probability curves shown in Fig. 11.3, the AAL ratio for the active year starting on 26/04/2017, (i.e., just after the tail of the 2016–2017 Central Italy sequence) was found to be three times higher than that for a random year and about four times higher compared to the estimate obtained using a conventional mainshock-only seismicity model. If the onset of the investigation time for the active period were moved to the middle of the 2016–2017 Central Italy sequence, the AAL estimates would have been a much higher multiple of the mainshock-only 3.4 per mille estimates of Fig. 11.14. Note that these AAL values are for the aggregated building inventory of the entire Umbria.

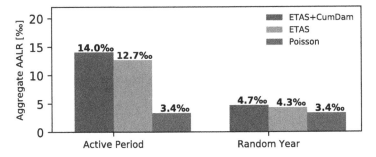

Fig. 11.14 AAL ratio for the entire building inventory of the Umbria region (Section 11.2) (as a percentage of the total replacement cost) computed using mainshock only seismicity and all seismicity for two distinct periods

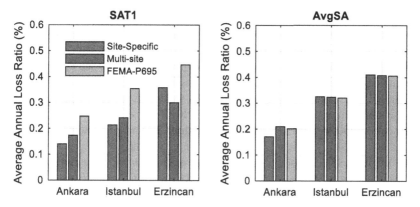

Fig. 11.15 Estimates of the AAL ratio for the same 7-story reinforced concrete building, whose vulnerability functions were derived based on SAT_1 (left) and $AvgSA$ (right) (Section 11.3). In blue the estimates obtained using site-hazard-consistent ground motion selection; in red the estimates computed using the regional-hazard-consistent ground motion selection proposed by Kohrangi et al. (2017b); in yellow the estimates from the application of the FEMA-P695 ground motion record set

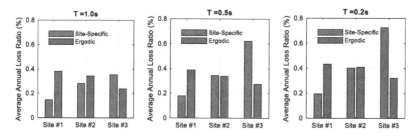

Fig. 11.16 Estimates of the AAL ratio for the three structures with the fundamental period of vibration of $T_1 = 0.2, 0.5, 1.0$ s located at the three sites (Section 11.4). In blue the estimates based on the site hazard computed via a site-specific non-ergodic GMPE; in red the estimates based on the site hazard computed via a traditional ergodic GMPE

Papadopoulos and Bazzurro (2021) pointed out, however, that this AAL increase due to the clustered seismicity is highly dependent on the vicinity of each building to the epicenters of the then-ongoing sequence. In fact, individual building AAL estimates were found amplified by more than one order of magnitude (compared to the Poissonian mainshock-only case) close to the epicenter of the 2016 Norcia earthquake, while they converge to the estimates for a random year as one moves away from it.

Figure 11.15 shows the estimates of the AAL ratio for the same 7-story reinforced concrete building located at sites in Ankara, Istanbul and Erzincan using SAT_1 (left) and $AvgSA$ (right) as the IM adopted for the development of the vulnerability function. The same considerations made for the mean annual rate of exceedance loss curves in the corresponding section hold here as well. The blue bars on the right panel of Fig. 11.15 computed using site-specific hazard-consistent ground motions and

AvgSA, a much superior and more sufficient IM, are to be considered as the closest estimates to the true but unknown values of the AAL. The traditional approach that entails developing vulnerability functions using SAT_1 as IM and sets of ground motions not necessarily consistent with the regional hazard is prone to causing bias of unknown sign (here positive, namely too conservative) and amplitude in the AAL estimates. If one wants for historical reasons to keep SAT_1 as the IM of choice, then the multi-site approach of Kohrangi et al. 2017b for selecting regionally hazard consistent ground motions is a practical procedure. The recommendation, however, as it appears clearly from an inspection of the right panel of Fig. 11.15, is to use that procedure anchored to *AvgSA* instead. It is remarkable, however, how the use of *AvgSA* decreases the importance of record selection. Even the random set of ground motions provided by FEMA-P695 yields excellent AAL estimates, at least for this example.

Finally, Fig. 11.16 displays the estimates of the AAL ratio for the three structures with the fundamental period of vibration of $T_1 = 0.2, 0.5, 1.0$ s located at the three sites for which partially non-ergodic GMPEs were developed. It is clear that only for Site #2 the estimates obtained using the traditional approach whose hazard is computed via ergodic GMPEs (red bars) are similar to the more precise ones that account for the specificity of the site for ground motion prediction. This occurs because the non-ergodic GMPEs and the ergodic GMPEs have similar median values for spectral accelerations in the neighborhoods of T_1 of the different structures. The characteristics of Sites #1 and #3 are significantly different than those of the average sites with same Vs_{30} whose recordings were used to develop the ergodic GMPE. Therefore, the traditional approach would overestimate by about 100% the AAL for all structures at Site #1 and underestimate the AAL at Site #3 by 40% for the $T_1 = 1.0$ s structure and by more than 100% the AAL of the two stiffer structures. Unfortunately, how much the peculiarities of each site deviate from those of the average site with same Vs_{30} is a piece of information seldom known a priori. Therefore, unless more data (enough recordings of past earthquakes in this case) at more sites becomes available the widespread use of more precise non-ergodic GMPEs will be limited to assessing hazard and risk at specific sites and its use in portfolio risk analysis will be prevented.

11.6   Conclusions and Recommandations

Catastrophe risk assessment models are at the core of many risk mitigation decisions made by a wide variety of stakeholders. The quality of these models and, consequently, the accuracy of the risk estimates they provide have steadily improved in the last 30 years since the first ones were developed. However, the current models used for commercial portfolio risk analyses have some known caveats that may lead to biased risk estimates. In this chapter we discussed two shortcomings existing in all current earthquake risk assessment models, namely the neglect of earthquake sequences and the use of region-generic ground motions for developing vulnerability functions.

Ignoring the spatial clustering of seismicity can induce a negative bias (i.e., one that leads to unconservative values) in loss estimates in a few different ways. Naturally, accounting for the earthquakes that are normally discarded during declustering yields higher AAL estimates. The AAL increase can arguably depend on several factors such as the seismicity conditions prior to the time when the analysis is conducted, the degree of seismicity clustering in a given region, the susceptibility of the building stock to damage accumulation, and the size and spatial distribution of the portfolio, to name a few. As an illustrative example, we presented the work of Papadopoulos and Bazzurro (2021), which compared AAL estimates for the Umbria region in Central Italy using different initial conditions, i.e. an active 1-year period as described earlier and a year with random initial conditions. The underestimation of the AAL, especially in periods of heightened seismicity, for all assets close to ongoing sequences is just too large to ignore.

Using insufficient ground motion IMs, such as SAT_1, and hazard-inconsistent ground motions for developing analytical vulnerability functions for classes of like-buildings are two features of the current practice that should be discontinued. The resulting vulnerability curves could be unbiased either positively or negative to such an extent unknown to the structural engineers who developed it. Practical workaround techniques that address both issues have been shown here to work well in removing most of such a bias. Their applicability in real life portfolio risk assessment is certainly recommended. Of the two suggested techniques, the use of $AvgSA$ as the pivotal IM for vulnerability functions development is the one that we expect to be adopted sooner, since it is both easier to grasp for engineers and the most effective of the two.

The use of non-ergodic GMPEs is certainly one of the future enhancements of earthquake risk assessment models for portfolios of structures. However, the road to arrive to destination still requires collecting more data and doing more research before its use can be recommended for this purpose. It must be noted, though, that partially non-ergodic GMPEs that account for specific characteristics of both seismogenic sources and seismic wave traveling paths and attenuation in certain regions are certainly closer to practical fruition. In all the regions where enough data is available, the use of partially non-ergodic GMPEs for hazard and risk assessment should certainly be considered.

References

Abrahamson NA, Silva WJ, Kamai R (2014) Summary of the ASK14 ground motion relation for active crustal regions. Earthq Spect 30:1025–1055

Akkar S, Sandıkkaya M, Şenyurt M, Sisi AA, Ay B, Traversa P, Douglas J, Cotton F, Luzi L, Hernandez B, Godey S (2014) Reference database for seismic ground-motion in Europe (RESORCE). Bull Earthq Eng 12:311–339

Akkar S, Sucuoglu H, Yakut A (2005) Displacement-based fragility functions for low- and mid-rise ordinary concrete buildings. Earthq Spect 21:901–927

Atik AL, Abrahamson N, Bommer JJ, Scherbaum F, Cotton F, Kuehn N (2010) The variability of ground-motion prediction models and its components. Seism Res Letters 81:794–801

Anderson JG, Brune JN (1999) Probabilistic seismic hazard analysis without the ergodic assumption. Seism Res Letters 70:19–28

ATC-13 (1985) Earthquake damage evaluation data for California. In: ATC-13, R. (ed.) Applied Technology Council. Redwood City, California, USA

Bal İE, Crowley H, Pinho R, Gülay FG (2008) Detailed assessment of structural characteristics of Turkish RC building stock for loss assessment models. Soil Dyn Earthq Eng 28:914–932

Baltay AS, Hanks TC, Abrahamson NA (2017) Uncertainty, variability, and earthquake physics in ground-motion prediction equations. Bull Seism Soc Am 107:1754–1772

Barbat A, Yépez Moya F, Canas J (1996) Damage scenarios simulation for seismic risk assessment in urban zones. Earthq Spect 12:371–394

Bard P-Y, Bora SS, Hollender F, Laurendeau A, Traversa P (2019) Are the Standard V S-Kappa host-to-target adjustments the only way to get consistent hard-rock ground motion prediction? Pure App Geoph, 1–20

Basili R, Brizuela B, Herrero A, Iqbal S, Lorito S, Maesano FE, Murphy S, Perfetti P, Romano F, Scala A (2019) NEAMTHM18 Documentation: the making of the TSUMAPS-NEAM tsunami hazard model 2018

Bianchini M, Diotallevi P, Baker J (2010) Prediction of inelastic structural response using an average of spectral accelerations. 10th International Conference on Structural Safety and Reliability (ICOSSAR09). Osaka, Japan

Bindi D, Kotha S-R, Weatherill G, Lanzano G, Luzi L, Cotton F (2018a) The pan-European engineering strong motion (ESM) flatfile: consistency check via residual analysis. Bull Earthq Eng, 1–20

Bindi D, Spallarossa D, Picozzi M, Scafidi D, Cotton F (2018b) Impact of magnitude selection on aleatory variability associated with ground-motion prediction equations: Part I—Local, energy, and moment magnitude calibration and stress-drop variability in central Italy. Bull Seism Soc Am 108:1427–1442

Bindi D, Kotha S (2020) Spectral decomposition of the engineering strong motion (ESM) flat file: regional attenuation, source scaling and Arias stress drop. Bull Earthq Eng, 1–26

Bindi D, Picozzi M, Spallarossa D, Cotton F, Kotha SR (2019) Impact of magnitude selection on aleatory variability associated with ground-motion prediction equations: part II—analysis of the between-event distribution in Central Italy. Bull Seism Soc Am 109:251–262

Bommer JJ, Crowley H (2006) The Influence of Ground-Motion Variability in Earthquake Loss Modelling. Bull Earthq Eng Des 4:231–248

Bommer JJ, Abrahamson NA (2006) Why do modern probabilistic seismic-hazard analyses often lead to increased hazard estimates? Bull Seism Soc Ame 96:1967–1977

Boore DM, Stewart JP, Seyhan E, Atkinson GM (2014) NGA-West2 equations for predicting PGA, PGV, and 5% damped PSA for shallow crustal earthquakes. Earthq Spect 30:1057–1085

Bradley BA (2010) A generalized conditional intensity measure approach and holistic ground-motion selection. Earthq Eng Struct Dyn 39:1321–1342

Brune JN (1970) Tectonic stress and the spectra of seismic shear waves from earthquakes. J Geoph Res 75:4997–5009

Brzev S, Scawthorn A, Charleson L, Allen M, Greene K, Jaiswal K, Silva V (2013) GEM Technical Report 2013–02. In: 2.0, G. B. T. V. (ed.). GEM Foundation, Pavia, Italy

Calvi G, Pinho R (2004) LESSLOSS. A European integrated project on risk mitigation for earthquakes and landslides. In: 2004/02, R. (ed.). European School for Advanced Studies in Reduction of Seismic Risk (ROSE School): Pavia, Italy

Calvi G, Pinho R, Magenes G, Bommer JJ, Restrepo-Vélez L, Crowley H (2006) Development of seismic vulnerability assessment methodologies over the past 30 years. ISET J Earthq Tech 43:75–104

Campbell KW, Bozorgnia Y (2014) NGA-West2 Ground motion model for the average horizontal components of PGA, PGV, and 5% damped linear acceleration response spectra. Earthq Spect 30:1087–1115

Chounet A, Vallée M, Causse M, Courboulex F (2018) Global catalog of earthquake rupture velocities shows anticorrelation between stress drop and rupture velocity. Tectonophysics 733:148–158

Cordova P, Deierlein G, Mehanny S, Cornell C (2000) Development of a two-parameter seismic intensity measure and probabilistic assessment procedure. In: The Second US-Japan Workshop on Performance-Based Earthquake Engineering Methodology for Reinforced Concrete Building Structures, September, 187–206

Cotton F, Archuleta R, Causse M (2013) What is sigma of the stress drop? Seismol Res Lett 84:42–48

D'ayala D, Meslem A, Vamvatsikos D, Porter K, Rossetto T (2014) Guidelines for analytical vulnerability assessment of low/mid-rise buildings. GEM Technical Report 2014–12: Global Earthquake Model Foundation, Pavia, Italy

Di Pasquale G, Orsini G, Romeo R (2005) New developments in seismic risk assessment in Italy. Bull Earthq Eng 3:101–128

Douglas J (2004) An investigation of analysis of variance as a tool for exploring regional differences in strong ground motions. J Seism 8:485–496

Douglas J (2014) Fifty years of ground-motion models. In: Second European Conference on Earthquake Engineering and Seismology (2ECEES): a joint event of the 15th European Conference on Earthquake engineering & 34th General Assembly of the European Seismological Commission

Eads L, Miranda E, Krawinkler H, Lignos DG (2013) An efficient method for estimating the collapse risk of structures in seismic regions. Earthq Eng Struct Dyn 42:25–41

Eads L, Miranda E, Lignos D (2015) Average spectral acceleration as an intensity measure for collapse risk assessment. Earthq Eng Struct Dyn

Faccioli E, Paolucci R, Vanini M (2015) Evaluation of probabilistic site-specific seismic-hazard methods and associated uncertainties, with applications in the Po Plain, northern Italy. Bull Seism Soc Am 105:2787–2807

FEMA-P58 (2012) Federal emergency management agency: Seismic performance assessment of buildings, prepared by the Applied Technology Council for the Federal Emergency Management Agency. In: AGENCY, F. E. M. (ed.). Washington, DC

FEMA-P695 (2009) Quantification of building seismic performance factors. 201 Redwood Shores Parkway, Suite 240, Redwood City, California 94065 Applied Technology Council

FEMA (2003) HAZUS-MH Technical Manual. In: AGENCY, F. E. M. (ed.). Washington, DC, USA

Field E, Porter K, Milner K (2017) A prototype operational earthquake loss model for California based on UCERF3-ETAS—A first look at valuation. Earthq Spect, 33

Gardner J, Knopoff L (1974) Is the sequence of earthquakes in southern California, with aftershocks removed, Poissonian. Bull Seism Soc Am 64:1363–1367

Haselton C, Baker J, Liel A, Deierlein G (2011) Accounting for ground-motion spectral shape characteristics in structural collapse assessment through an adjustment for epsilon. J Struct Eng 137:332–344

Jalayer F (2003) Direct probabilistic seismic analysis: implementing non-linear dynamic assessment. PhD Thesis, Stanford University

Jalayer F, Ebrahimian H (2016) Seismic risk assessment considering cumulative damage due to aftershocks. Earthq Eng Struct Dyn

Jayaram N, Lin T, Baker J (2011) A computationally efficient ground-motion selection algorithm for matching a target response spectrum mean and variance. Earthq Spect 27:797–815

Jayaram N, Shome N, Rahnama M (2012) Development of earthquake vunerability functions for tall buildings. Earthq Eng Struct Dyn 41:1495–1514

Kale Ö, Akkar S, Ansari A, Hamzehloo H (2015) A ground-motion predictive model for Iran and Turkey for horizontal PGA, PGV, and 5% damped response spectrum: Investigation of possible regional effects. Bull Seism Soc Am 105:963–980

Kappos A, Pitilakis K, Stylianidis K (1995) Cost-benefit analysis for the seismic rehabilitation of buildings in Thessaloniki, based on a hybrid method of vulnerability assessment. In: Proceedings of the Fifth International Conference on Seismic Zonation, Nice, France, 406–413

Kappos A, Stylianidis K, Pitilakis K (1998) Development of seismic risk scenarios based on a hybrid method of vulnerability assessment. Nat Hazards 17:177–192

Kazantzi A, Vamvatsikos D, Porter K (2015) Analytical seismic vulnerability assessment for a class of modern low-rise steel frames. In: 12th International Conference on Applications of Statistics and Probability in Civil Engineering, ICASP12. Vancouver, Canada

Kennedy R, Ravindra M (1984) Seismic fragilities for nuclear power plant risk studies. Eng Design 79:47–68

Kohrangi M, Bazzurro P, Vamvatsikos D (2016) Vector and scalar IMs in structural response estimation, Part II: Building Demand Assessment. Earthq Spect 32:1525–1543

Kohrangi M, Bazzurro P, Vamvatsikos D, Spillatura A (2017) Conditional spectrum-based ground motion record selection using average spectral acceleration. Earthq Eng Struct Dyn 46:1667–1685

Kohrangi M, Kotha SR, Bazzurro P (2020) Impact of partially non-ergodic site-specific probabilistic seismic hazard on risk assessment of single buildings. Earthq Spect (in-review)

Kohrangi M, Vamvatsikos D, Bazzurro P (2016) Implications of intensity measure selection for seismic loss assessment of 3-D buildings. Earthq Spect 32:2167–2189

Kohrangi M, Vamvatsikos D, Bazzurro P (2017) Site dependence and record selection schemes for building fragility and regional loss assessment. Earthq Eng Struct Dyn 46:1625–1643

Kotha SR, Bindi D, Cotton F (2016) Partially non-ergodic region specific GMPE for Europe and Middle-East. Bull Earthq Eng 14:1245–1263

Kotha SR, Bindi D, Cotton F (2017) From ergodic to region- and site-specific probabilistic seismic hazard assessment: method development and application at European and Middle Eastern sites. Earthq Spect 33:1433–1453

Kotha SR, Weatherill G, Bindi D, Cotton F (2020) A regionally adaptable ground-motion model for shallow crustal earthquakes in Europe. Bull Earthq Eng

Landwehr N, Kuehn NM, Scheffer T, Abrahamson N (2016) A nonergodic ground-motion model for california with spatially varying coefficients. Bull Seism Soc Am 106:2574–2583

Lanzano G, Sgobba S, Luzi L, Puglia R, Pacor F, Felicetta C, D'amico M, Cotton F, Bindi D (2018) The pan-European engineering strong motion (ESM) flatfile: compilation criteria and data statistics. Bull Earthq Eng, 1–22

Lin P-S, Chiou B, Abrahamson N, Walling M, Lee C-T, Cheng C-T (2011) Repeatable source, site, and path effects on the standard deviation for empirical ground-motion prediction models. Bull Seism Soc Am 101:2281–2295

Luco N, Cornell C (2007) Structure-specific scalar intensity measures for near-source and ordinary earthquake ground motions. Earthq Spect 23:357–392

Luen B, Stark P (2011) Are declustered earthquake catalogs Poisson. Statistics, 1–8

Marzocchi W, Taroni M (2014) Some thoughts on declustering in probabilistic seismic-hazard analysis. Bull Seism Soc Am 104:1838–1845

Mcguire RK (2004) Seismic hazard and risk analysis. Earthquake Engineering Research Institute

Mitrani-Reiser J (2007) An ounce of prediction: probabilistic loss estimation for performance-based earthquake engineering. PhD Dissertation, Calofornia Institute of Technology

Noh H, Lallemant D, Kiremidjian A (2015) Development of empirical and analytical fragility functions using kernel smoothing methods. Earthq Eng Struct Dyn 44:1163–1180

Ogata Y (1998) Space-time point-process models for earthquake occurrences. Ann Inst Stat Math 50:379–402

Orsini G (1999) A model for buildings' vulnerability assessment using the parameterless scale of seismic intensity (PSI). Earthq Spect 15:463–483

Papadopoulos A, Bazzurro P (2021) Exploring probabilistic seismic risk assessment accounting for seismicity clustering and damage accumulation: Part II. Risk Analysis, Earthq Spect 37(1):386–408

Pinho R, Bommer J, Glaister SA (2002) Simplified approach to displacement-based earthquake loss estimation analysis. In: Proceedings of the 12th European Conference on Earthquake Engineering, London, UK: Paper No. 738 (on CD)

Porter K, Farokhnia K, Vamvatsikos D, Cho I (2014) Guidelines for component-based analytical vulnerability assessment of buildings and nonstructural elements. GEM Technical Report 2014–13: Global Earthquake Model Foundation, Pavia, Italy

Porter K, Kennedy R, Bachman R (2007) Creating fragility functions for performance-based earthquake engineering. Earthq Spect 23:471–489

Porter K, Kiremidjian A, Legrue J (2001) Assembly-based vulnerability of buildings and its use in performance evaluation. Earthq Spect 17:290–312

Radiguet M, Cotton F, Manighetti I, Campillo M, Douglas J (2009) Dependency of near-field ground motions on the structural maturity of the ruptured faults. Bulle Seism Soc Am 99:2572–2581

Restrepo-Vélez L, Magenes G (2004) Simplified procedure for the seismic risk assessment of unreinforced masonry buildings. In: Proceedings of the 13th World Conference on Earthquake Engineering, Vancouver, Canada

Rodriguez-Marek A, Cotton F, Abrahamson NA, Akkar S, Al Atik L, Edwards B, Montalva GA, Dawood HM (2013) A model for single-station standard deviation using data from various tectonic regions. Bull Seism Soc Am 103:3149–3163

Rossetto T, Elnashai A (2003) Derivation of vulnerability functions for european-type RC structures based on observational data. Eng Struct 25:1241–1263

Rossetto T, Ioannou I, Grant D, Maqsood T (2014) Guidelines for empirical vulnerability assessment. GEM Technical Report, GEM Foundation, Pavia

Sextos A, Risi R, Pagliaroli A, Foti S, Passeri F, Ausilio E, Cairo R, Capatti MC, Chiabrando F, Chiaradonna A, Dashti S, De Silva F, Dezi F, Durante M, Giallini S, Lanzo G, Sica S, Simonelli AL, Zimmaro P (2018) Local site effects and incremental damage of buildings during the 2016 Central Italy Earthquake sequence. Earthq Spect 34

Shokrabadi M, Burton HV (2017) Building service life economic loss assessment under sequential seismic events. Earthq Eng Struct Dyn 47

Shokrabadi M, Burton HV (2019) Regional short-term and long-term risk and loss assessment under sequential seismic events. Eng Struct 185:366–376

Silva V, Crowley H, Varum H, Pinho R, Sousa L (2015) Investigation of the characteristics of Portuguese regular moment-frame RC buildings and development of a vulnerability model. Bull Earthq Eng 13:1455–1490

Silva V, Crowley H, Varum H, Pinho R, Sousa R (2014) Evaluation of analytical methodologies used to derive vulnerability functions. Earthq Eng Struct Dyn 43:181–204

Sousa M, Costa AC, Carvalho A, Coelho E (2004) An automatic seismic scenario loss methodology integrated on a geographic information system. In: Proceedings of the 13th World Conference on Earthquake Engineering, Vancouver, Canada. Paper No. 2526 (on CD)

Stewart J, Lanzo G, Ausilio E, Cairo R, Bozzoni, F, Capatti MC, Pasqua F, Dezi F, Di Sarno L, Durante M, Simonelli AL, Foti S, Chiabrando F, Dabove P, Di Pietra V, Maschio P, Passeri F, Sgobio A, Teppati Losè L, Zimmaro P (2017a) Engineering reconnaissance following the October 2016 Central Italy Earthquakes

Stewart JP, Afshari K, Goulet CA (2017) Non-ergodic site response in seismic hazard analysis. Earthq Spect 33:1385–1414

Stiphout T, Schorlemmer D, Wiemer S (2011) The effect of uncertainties on estimates of background seismicity rate. Bull Seism Soc Am 101:482–494

Straub D, Der Kiureghian A (2008) Improved seismic fragility modeling from empirical data. Struct Safety 30:320–336

Tothong P, Luco N (2007) Probabilistic seismic demand analysis using advanced ground motion intensity measures. Earthq Eng Struct Dyn 36:1837–1860

Vamvatsikos D, Cornell C (2002) Incremental dynamic analysis. Earthq Eng Struct Dyn 31:491–514

Vamvatsikos D, Cornell C (2005) Developing efficient scalar and vector intensity measures for IDA capacity estimation by incorporating elastic spectral shape information. Earthq Eng Struct Dyn 34:1573–1600

Villar-Vega M, Silva V, Crowley H, Yepes C, Tarque N, Acevedo AB, Hube MA, Gustavo CD, María HS (2017) Development of a fragility model for the residential building stock in South America. Earthq Spect 33:581–604

Walling MA (2009) Non-ergodic probabilistic seismic hazard analysis and spatial simulation of variation in ground motion. University of California, Berkeley

Weatherill G, Kotha SR, Cotton F (2020) A Regionally-adaptable "scaled-backbone" ground motion logic tree for shallow seismicity in Europe: application in the 2020 European seismic hazard model. Bull Earthq Eng

Weatherill G, Pagani M, Garcia J (2016) Exploring earthquake databases for the creation of magnitude-homogeneous catalogues: tools for application on a regional and global scale. Geophysical J Int 206:1652–1676

Woessner J, Laurentiu D, Giardini D, Crowley H, Cotton F, Grünthal G, Valensise G, Arvidsson R, Basili R, Demircioglu MB, Hiemer S, Meletti C, Musson RW, Rovida AN, Sesetyan K, Stucchi M (2015) The 2013 European seismic hazard model: key components and results. Bull Earthq Eng 13:3553–3596

Zhang L, Werner M, Goda K (2018) Spatiotemporal seismic hazard and risk assessment of aftershocks of M9 Megathrust Earthquakes. Bull Seism Soc Am 108:3313–3335

Part III
Earthquake Insurance for Resilience

Chapter 12
The Role of Earthquake Insurance in Earthquake Risk Reduction and Resilience Building

Fouad Bendimerad

Abstract Resilience is defined as *"The ability to **prepare**and plan for, **absorb**, recover from and more successfully **adapt**to adverse events"* (US National Academies). Resilience has four pillars: • **Anticipate**: the ability to anticipate and reduce the impact of shocks through preparedness and planning, • **Absorb** the ability to absorb and cope with the impacts of shocks and stresses. • **Adapt**: the ability to change in response to multiple, long-term and future risks, and to learn and adjust after a shock materializes. • **Transform:** the ability to take deliberate steps to change the systems that create risk, vulnerability and or inequality. How does insurance intervene in building resilience? The outcome of insurance is to restore property and livelihoods in case of an adverse effect. It does that by providing a cash infusion into the socio-economic system of the affected communities immediately after the event. The cash is used to restore property and avoid interruption of commercial and industrial activity. Insurance also intervenes in terms of reducing impact of stresses (which are the more extensive types of risk) since it enables a system of "maintenance" by providing funds for recovery under minor but more frequent events. For most developing countries, governments have been the insurer of last resort when it comes to catastrophe risk (referred to as Cat Risk in the insurance industry). The reason is that level of cat insurance penetration in most developing countries is very low, sometimes lower than 1%. The assurance of government intervention coupled with the lack of effectiveness of the financial transaction associated with a traditional insurance policy negate any incentive for individuals to acquire a cat insurance policy. The Turkish Compulsory Insurance Program or TCIP is one of the early experiment to change that paradigm and to provide a meaningful role for cat insurance in emerging economies. After a slow start, TCIP has now developed the financial capacity and the spread of coverage to play a significant role both in the financing of risk but also in supporting earthquake risk reduction in Turkey. New cat insurance products based on parametric indexing have since emerged. These insurance products could further improve the efficiency of TCIP and other cat insurance pools by making them more attractive to individuals, thereby scaling up their contribution to building resilience.

F. Bendimerad (✉)
EMI, 3F Puno Bldg. Annex 47, Kalayaan Avenue, Diliman, 1101 Quezon City, Philippines
e-mail: fouadb@emi-megacities.org

© The Author(s) 2021
S. Akkar et al. (eds.), *Advances in Assessment and Modeling of Earthquake Loss*,
Springer Tracts in Civil Engineering,
https://doi.org/10.1007/978-3-030-68813-4_12

12.1 Resilience and System Theory

Resilience finds its theoretical basis in system theory. This is illustrated in Fig. 12.1. An urban system, in this case the city of Istanbul is a dynamic system subject to the five elements indicated in the figure. A system is resilient when negative environmental shocks and stresses cause the minimum impact to its outputs and will require minimum, if any external resources to maintain its positively productive steady state. An important point is that the resilience of a system is only as good as its weakest subsystem.

The value of resilience in any urban system (e.g., housing, transport, land use, etc.) is revealed by the risk diagram of Fig. 12.2, which shows what happens when insufficient resilience is designed into such system. Should a "system", prove to have too little resilience, it can go into a state of emergency even when a simple perturbation occurs. Conversely, should the initial state of the system has the necessary and sufficient resilience built-in from the start, then it should be able to sustain a "perturbation" and still maintain functionality, albeit at a some diminished mode of resilient operation. This means that a state of emergency within an urban system could potentially be averted should its initial design provide an appropriate level of resiliency.

In addition, the duration of the degraded operations would ostensibly be shorter with fewer critical systems being impacted. Finally, after the crisis comes to an end, the resilient systems could be built back safer using lessons learned. The new "normal" state would represent an even more resilient environment. We call this type of system "resilient risk-tolerant", as opposed to "unaware risk-averse". Risk-averse systems are those that cannot or will not accept that there are always risks involved; whereas resilient risk-tolerant ones are those that assess the risk and manage it by employing appropriate levels of resilience.

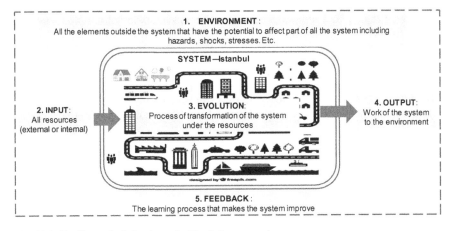

Fig. 12.1 Resilience finds its theoretical basis in system theory

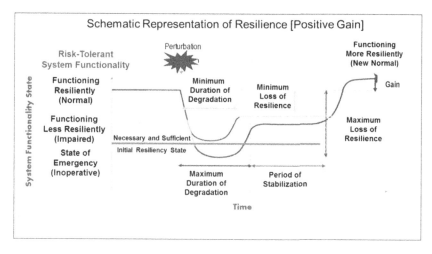

Fig. 12.2 Schematic representation of resilience indicating positive gain

Government can introduce several mechanisms to reduce risk. These may include structural investments, improve emergency response, as well as non-structural investments for awareness raising and capacity building. However, independently on how much investment society makes in reducing its risk, there is always some amount of risk that remains, called Residual Risk. This is where insurance intervenes. It offers the financial mechanism to finance residual risk. In the hypothetical example of Fig. 12.3, it is assumed that the combined outcome of all structural and non-structural interventions of government and non-government entities has resulted in a 65% reduction in the potential losses from a disaster. The 35% risk that remains is the residual risk that can ideally be financed through insurance (Fig. 12.3).

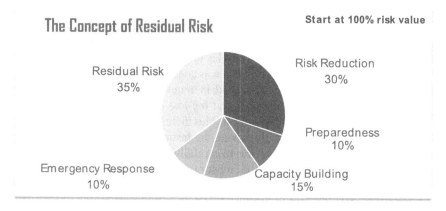

Fig. 12.3 Hypothetical case illustrating the residual risk concept

12.2 Insurance and Resilience

Within this context, how does insurance intervene in building resilience?

The ultimate outcome of insurance is to contribute to the restoration of property and livelihoods in the case of an adverse effect on a system. It does that by providing a cash infusion into the system immediately after the event. The cash is used to restore property and avoid interruption of commercial and industrial activity. In reference to Fig. 12.1, Insurance is part of the Input of the system. It is essentially a resource to the system. The big advantage of insurance as a resource is that it is an immediate cash infusion. In fact, the faster the insurance funds are put back into the system, the more resilience capital the system has. This corollary indicates that parametric approaches to Cat Insurance are favored over indemnity approaches since, in principle the distribution of funds is much faster in the former than the latter.

Insurance also intervenes in terms of reducing impact of stresses (which are the more extensive types of risk) since it enables a system of "maintenance" by providing funds for recovery under minor but more frequent shocks.

Finally, a third form of insurance intervention is through awareness raising about hazards and risks. A well-informed community or a business has confidence in the viability of the system itself. For example, by insuring its asset, a business reduces its liability and is in better financial position to increase investment and productivity.

This of course, describes an ideal insurance governance system built on trust, transparency, and effectiveness. However, as explained below, there have been barriers to this "perfect" system in the past.

12.3 How Does Cat Insurance Work?

Cat insurance deals with providing insurance coverage against natural disasters such as earthquakes, floods, extreme winds, and others (Bendimerad and Home 1999). It is a specific business line of the insurance industry and has particular rules. The operations and functions of cat insurance are often not understood by the public. One of the main reasons, is that Cat Insurance is complex to underwrite and manage by insurers and regulators alike. Until recently, cat risk was considered as a "non-insurable" risk because of the unpredictability of losses (both in frequency and severity). The concern of both the insurer and the regulator is that big-loss events (high frequency—low severity events) will wipe out the insurer surplus thereby endangering compensation of the policyholders and potentially driving the insurer into bankruptcy. To reduce such a risk, the primary insurance tends to carefully control its exposure to cat risk, while at the same time seeking to cede most of the risk to one or more reinsurers. There are several forms of contracts between a primary insurer and the reinsurer, but they essentially have the same objective, which is to transfer a portion of the risk from the insurance to the reinsurer for a share of the premium.

In the mid-nineties, catastrophe bonds (cat bonds) appeared on the capital market. These bonds enable both insurers and reinsurers to cede a portion of the risk directly to institutional investors on the capital market. Cat bonds offered a new way for the insurance industry to increase access to capital, thus improving its capacity to underwrite cat risk. Cat bonds proved to be of great interest to several institutional investors because they are not correlated to traditional investments such as the corporate stocks, corporate or government bonds, real estate derivatives and others. They offer investors an attractive option for portfolio diversification (Canabarro et al. 2000).

12.4 Why Does Insurance Matter in Building Resilience?

Insurance is vital to the economy of a country. For most countries, governments have been the insurer of last resort when it comes to cat risk. The reason is that level of cat insurance penetration in many countries is very low. The assurance of government intervention coupled with the lack of effectiveness of the financial transaction negated any incentive for individuals to acquire a cat insurance policy. Other elements were also at play in affecting the business case for earthquake insurance:

– The fact that the events are rare reduced the public awareness (i.e., it will not happen in my lifetime)
– The perception that I am fine because my property/business was not affected in the last event. It only happened to others
– The perception that construction codes provided "disaster-proof" structures
– A general 'chronic" awareness about cat insurance and cat risk among the general public
– High cost—For a large segment of society the cost of cat insurance is not affordable.

The introduction of cat models in the early nineties revolutionized the Cat insurance business by providing physical approaches to modeling losses. Actuarial models, which were used by the insurance industry, were very poor indicators of future losses since cat events are rare events. The historical data is too sparse to yield reliable actuarial models making both pricing and reinsurance transactions highly uncertain. Cat models provided greater capabilities to link the physical characteristics of natural hazard events to the vulnerability of the exposed assets, thereby projecting future losses with greater accuracy, while enabling insurers to understand the spatial distribution of exposure and loss along the insurer's lines of business. Cat models provided a means to better anticipate the future and to quantify volatility. The cat models provided the tool to more accurately price cat risk and more efficiently transact contracts between all parties involved, particularly the insurer, reinsurer and investor. Cat models made new financial products such as Cat bond possible and opened the door for more effective financial transactions embedded in the indexed and parametric insurance products. These models significantly reduce the cost of an insurance transaction compared to the conventional indemnity practice. Further, they increase capacity for reinsurance while at the same time providing more peace of mind to the insured. It is these latest advances that have made insurance an essential parameter in resilience building.

12.5   The New Dynamic in Cat Risk Financing

Two successive early events were wake up calls for the insurance industry in its handling of cat risks: 1992 Hurricane Andrew and 1995 Northridge earthquake. These events demonstrated that extreme events could create devastating losses to society, and consequently to insurers. The immediate impact was in the universal shift of the insurance industry towards the use of cat models. Since then many catastrophic events have taken place around the globe and the cumulative losses have staggered prompting a rise in awareness among governments and the public, that new approaches for managing extreme events from natural hazards were needed.

New phenomena are also taking place that caused the increase awareness and the call for more action not only by the insurance industry but also by government.

- Urbanization is causing greater accumulation of assets in cities and urban agglomerations;
- Wealth accumulation and increases in the standards of living have increased the value of assets;
- Increase in the severity (and arguably frequency) of extreme weather events due climate change;
- Domino-effect due to the complexity of the urban environment that can multiply the losses.

Governments are finding it more and more costly to come after a disaster and pick up the bill. They are looking at ways to not only increase the role of insurance, but to incorporate risk financing as part of the tool set of financial instruments to reduce contingent liabilities, reduce the financial impact and improve recovery from disasters. They are also realizing that "physical protection" through infrastructure is not necessary full protection. Typhoon Hagibis of October 2019 in Japan caused more than 100 dikes to fail, at the time where the Japanese authorities thought that had controlled flood risk in the country through major multi-decades flood protection investments. In addition to strengthening physical infrastructure, countries are looking for new strategies to cover losses and to shift these losses from government to individuals, insurers and to the capital market.

12.6 TCIP as an Early Experiment

TCIP is an early experiment aimed at increasing penetration by making cat insurance mandatory for homeowners in Turkey. This approach can be considered to represent the very first "line of defense" for government to reduce its contingent liability by shifting losses to the property owners and the insurance industry. The basic concept of insurance of spreading risk widely among as many policyholders as possible is applied, thereby reducing premium cost and making insurance affordable. Certain mechanisms are used to ensure that the property owner is obligated to contract insurance (e.g., linking insurance to real-estate transactions). While in its early stages, TCIP struggled to accomplish its goals, it has progressively build strength mostly through sound insurance practices, awareness raising, keeping affordability, and relying on scientific modeling. In 2016, TCIP has 136 billion Euro coverage capacity for earthquake losses in Turkey. A true evidence of resilience contribution to Istanbul and other major cities in Turkey since that capital can put back into the economy immediately after an earthquake with essentially zero cost to the government. Another significant advantage is that TCIP would put back its contribution to the regional and local economies since the direct impacts of an earthquake event are local and regional. The TCIP coverage capacity is a true evidence of resilience contribution to Istanbul, other major cities in Turkey and the country as a whole.

12.7 More Innovation in the Market

With the support of the World Bank and other international financial institutions (IFIs), more innovation is coming to the financing of disaster risk, by introducing sophisticated strategies and solutions aimed at reducing the transaction cost and building a greater demand. The underlying objective remains the protection of the state finances, individual assets, and the economy through sustainable and efficient risk financing mechanisms that cover projected disaster-related expenditures and at the same time, reduce volatility and increase predictability. Greater demand is accomplished through an increase in transparency and efficiency. The use of catastrophe loss models makes this achievable by providing scientifically auditable projections of losses and a rigorous mapping of characteristics and values of the exposed assets. The following provide a few examples of innovative disaster risk financing mechanisms that are aimed not only at providing liquidity after a disaster but also building resilience.

12.7.1   Indonesia: Pooling Fund Untuk Bencana—PFB

Core to the government strategies for risk financing, is the establishment of the Pooling Fund for Disaster (*Pooling Fund untuk Bencana*—PFB) (World Bank 2020). The PFB is a dedicated disaster reserve fund (a 'pooling fund') aimed at strengthening the financial and fiscal resilience of the Government of Indonesia to natural disasters and health-related shocks. The pooling fund is expected to help improve disaster financing by:

(i)   Enabling the government to accrue unspent budget allocations for disaster response to save for future years and build reserves;

(ii)   Improving efficiency in the use of funds from the state budget for post-disaster expenditures through improved up front planning and budgeting;

(iii)   Leveraging additional financial instruments by linking them directly to the pooling funds with clear and pre-agreed operating procedures (e.g. development partner contingent financing, market based risk transfer, or international assistance following events);

(iv)   Connecting the pooling fund to clear pre-arranged disbursement channels and rules, thereby increased speed and transparency of post-disaster spending and providing predictability to implementing agencies on the availability of funds; and

(v)   Increasing the ability to link risk financing to incentives and investment for all phases of disaster risk management, including preparedness and prevention.

While the PFB is still at the definition stage, it has secured a $510 million funding commitment from the World Bank and is well embedded in the overarching government reforms on building fiscal, physical, and social resilience, which are outlined in the recently approved National MediumTerm Development Plan (RPJMN) for 2020–2024.

12.7.2   Philippine: The Philippine City Disaster Insurance Pool (PCDIP)

The Philippine City Disaster Insurance Pool (PCDIP) is a disaster risk financing (DRF) solution to enable immediate access to early recovery funding for cities in the Philippines (Asia Development Bank 2018). The pool is being developed by the Government of the Philippines with support from the Asian Development Bank. It is one element of the Government of the Philippines strategy to create catastrophe financial self-sufficiency of local government units—LGUs—(i.e., provinces and cities). Its innovation is that it brings LGUs together into a pool to create a DRF scheme that gives them quick access to funding for immediate emergency response action and early recovery. The PCDIP would provide immediate short term liquidity

Fig. 12.4 Disaster risk financing instruments for cities (Philippine)

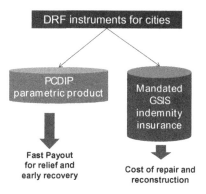

to city and provincial governments for the relief and early recovery phases of a disaster. A simple schematic on the functioning of the PCDIP is indicated in Fig. 12.4.

Earlier studies and recent events indicated that local governments that had cash in hands at the time of a disaster event, were able to recover faster than those who had to wait for an allocation from the central government budget. The PCDIP would complement existing disaster risk financing instruments including the Government Services Insurance System (GSIS) and would serve as an additional source of near-immediate liquidity in the event of a disaster. Initially, PCDIP covers two hazards: Tropical cyclone (wind speed) and earthquake (ground shaking). For each hazard cities can choose the level of premium that would pay for the combination of minimum and maximum payouts they would receive. These parameters are anchored on a modeled index parametric trigger structure. Recognizing that cities have different hazard levels and varying financial capacity, the design of the PCDIP attempts to provide a flexible, accessible and valuable insurance scheme; it offers certainty and speed in funding for pre-determined triggering events. Cities will know within days of a catastrophe event occurring what sum they will receive and then physically receive the cash shortly after. The PCDIP has not seen the day yet, but as a concept it is innovative and addresses a unique constituencies (i.e., local authorities) that are at the front line of disaster response and recovery.

12.8   Conclusions

Insurance and risk financing are important elements in building resilience. The overarching mission is to protect state finances, public and private assets and the economy of a country. Cat insurance is often ineffective in developing and emerging economies because it is not adequately regulated and managed and does not take advantage of the scientific knowledge offered by catastrophe loss models. Governments must look at cat insurance in the overall scheme of financing disaster risk and building resilience. This may require reforms in the insurance regulatory system that enables more innovation and reduces morale hazard. Through these steps, governments can more effectively deal with the escalating losses from disasters and reduce its contingent liability by shifting cost to individuals, insurance and the capital market. Recent examples show that innovation is bringing sustainable, transparent and efficient risk financing mechanisms at a lower cost than traditional insurance.

References

Asia Development Bank (2018). Philippine City Disaster Insurance Pool (PCDIP): Rationale and Design.

Bendimerad F and Hom S (1999). CATASTROPHE Modeling, Risk Management, 46(5):26.

Canabarro E, Finkemeier M, Anderson RR and Bendimerad F (2000). Analyzing insurance-linked securities, The Journal of Risk Finance, 1(2): 49–75.

World Bank (2020). Indonesia Disaster Risk Finance & Insurance (P173249), Project Information Document (https://documents1.worldbank.org/curated/en/225361591241320010/pdf/Concept-Project-Information-Document-PID-Indonesia-Disaster-Risk-Finance-Insurance-P173249.pdf)

Chapter 13
Fire Following Earthquake—The Potential in Istanbul

C. Scawthorn

Abstract Fire following earthquake is a little recognized risk in seismic regions with significant wood building inventories. Methods exist for quantifying this risk, and examples are provided in this chapter for San Francisco, Istanbul and Montreal. There are many opportunities for reducing this risk, and examples are provided regarding reducing fire station vulnerability and improving emergency firefighting water supply. Once accomplished however, vigilance is required to maintain these mitigation measures.

13.1 Introduction

Fire following earthquake refers to a series of events or a *stochastic process* initiated by a large earthquake. Fires occur following all earthquakes that significantly shake a human settlement but are generally only a very significant problem in a large metropolitan area predominantly comprised of densely spaced buildings. In such circumstances, the multiple simultaneous ignitions can lead to catastrophic conflagrations that by far are the dominant agent of damage for that event. Regions of high seismicity with large metropolitan areas predominantly comprised of densely spaced wood buildings are particularly at risk, and include Japan, New Zealand, parts of Europe and western North America. Istanbul is a major metropolitan area subject to large earthquakes that has historically sustained large conflagrations, so that the potential for fire following earthquake losses is a significant concern and the focus of this chapter.

C. Scawthorn (✉)
Pacific Earthquake Engineering Research Center, University of California at Berkeley, Berkeley, CA 94720-1792, USA
e-mail: cscawthorn@sparisk.com

© The Author(s) 2021
S. Akkar et al. (eds.), *Advances in Assessment and Modeling of Earthquake Loss*,
Springer Tracts in Civil Engineering,
https://doi.org/10.1007/978-3-030-68813-4_13

287

Large fires, for example measured in terms of square kilometers of burnt area, have not been unique to fires following earthquakes—indeed, the great fires of London (1666) and Chicago (1871) are only the most noteworthy of a long succession of non-earthquake related urban peace-time conflagrations. Istanbul in its long history has had a number of large conflagrations:

> ... the terrible earthquake which took place in 1509, known as "the Little Day of Judgment", dealt a terrible blow to the city's brick and stone houses. Fearing the effects of further earthquakes people began to build their houses of wood instead and in a short space of time Istanbul became a city of wooden houses. ... This inevitably led to a rapid increase in the number of fires in the city. As well as the earthquakes of 1765 and 1894, which also caused terrible destruction, Istanbul's greatest enemy has always been fire. ...Sometimes these fires would start from the banks of the Golden Horn and burn until they reached Aksaray or even the Sea of Marmara. The last great fires of Istanbul were the Hocapaşa fire of 1865, the Beyoglu fire of 1870, the Laleli fire of 1911, the Gedikpaşa fire of 1912 and the Cibali-Fatih-Altınmermer fire of 1918. ...The last big fire of this kind destroyed a large part of the Fener district in 1941 (Ansal 2003).

Still, the two largest peace-time urban conflagrations in history have been fires following earthquakes—1906 San Francisco and 1923 Tokyo, the latter resulting in the great majority of the 140,000 fatalities.

Although a combination of a professionalized fire service, improved water supply and better building practices has largely eliminated non-earthquake related large urban conflagrations in Istanbul, there is still a gap—an Achilles Heel—which is fire following earthquake. This is due to the correlated effects of a large earthquake simultaneously causing numerous ignitions, degrading building fire resistive features, dropping pressure in water supply mains, saturating communications and transportation routes, and thus allowing some fires to quickly grow into conflagrations that outstrip local resources. It is not sufficiently appreciated that the key to modern fire protection is a well-drilled rapid response by professional firefighters in the early stages of structural fires, arriving in time to suppress the fires while that is still relatively feasible. A typical response goal for urban fire departments for example is 4 min from time of report to arrival. If suppression is delayed, due either to delayed response, or lack of water, a single structural fire can quickly spread to neighboring buildings and grow to the point where an entire municipalities' fire resources are required, and perhaps even assistance from neighboring communities. This is for a single ignition. Simply put, most fire departments are not sized or equipped to cope with the fires following a major earthquake. A major earthquake and its associated fires is a low probability event for which, although having very high potential consequences, it may not be feasible to adequately prepare.

13.2   Analysis of Fire Following Earthquake

The first step towards solving any problem is analyzing the problem and quantifying its effects. A full probabilistic methodology for analysis of fire following earthquake was developed in the late 1970s (Scawthorn et al. 1981) and applied to major cities in western North America (Scawthorn and Khater 1992), Japan and Istanbul, the methods for which are detailed in (Scawthorn et al. 2005), so that only a brief review is presented here. In summary, the steps in the process are:

- *Occurrence of the earthquake*—causing damage to buildings and contents, even if the damage is as simple as knockings things (such as candles or lamps) over.
- *Ignition*—whether a structure has been damaged or not, ignitions will occur due to earthquakes. The sources of ignitions are numerous, ranging from overturned heat sources, to abraded and shorted electrical wiring, to spilled chemicals having exothermic reactions, to friction of things rubbing together.
- *Discovery*—at some point, the fire resulting from the ignition will be discovered, if it has not self-extinguished (this aspect is discussed further, below). In the confusion following an earthquake, the discovery may take longer than it might otherwise.
- *Report*—if it is not possible for the person or persons discovering the fire to immediately extinguish it, fire department response will be required. For the fire department to respond, a Report to the fire department has to be made. Communications system dysfunction and saturation will delay many reports.
- *Response*—the fire department then has to respond, but is impeded by non-fire damage emergencies they may have to respond to (e.g., building collapse) as well as transportation disruptions.
- *Suppression*—the fire department then has to suppress the fire. If the fire department is successful, they move on to the next incident. If the fire department is not successful, they continue to attempt to control the fire, but it spreads and becomes a conflagration. Success or failure hinges on numerous factors including water supply functionality, building construction and density, wind and humidity conditions, etc. If unable to contain the fire, the process ends when the fuel is exhausted or when the fire comes to a firebreak.

This process is also shown in Fig. 13.1 which is a Fire Department Operations Timeline. Time is of the essence for the fire following earthquake problem. In this figure, the horizontal axis is Time, beginning at the time of the earthquake, while the vertical axis presents a series of horizontal bars of varying width. Each of these bars depicts the development of one fire, from ignition through growth or increasing size (size is indicated by the width or number of bars). Fire following earthquake is a highly non-linear process, modeling of which does not have great precision and is such that in many cases the only clear result is differentiation between situations of a few small fires, versus major conflagration.

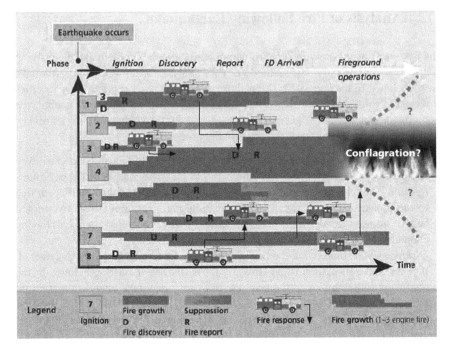

Fig. 13.1 Fire following earthquake timeline

13.2.1 Assets at Risk and Ignitions

Ignitions are a function of shaking intensity and the total amount of building floor area exposed to that shaking—the more buildings, the more sources of ignitions. Figure 13.2 shows total building floor area (in sq. m., per 0.01° grid cell) for Istanbul. While Istanbul was historically composed primarily of wood buildings, in recent decades the region's building stock has become almost entirely masonry and reinforced concrete—good for fire, questionable for earthquake.

Figure 13.3 shows estimated shaking intensity for a M7.5 earthquake expected to occur on the North Anatolian fault in close proximity to Istanbul (Ansal 2003). The actual number of ignitions varies with each ground shaking scenario—for the M7.5 event most of the Study Area would be subjected to PGAs of 0.1—0.4 g, meaning that the total number of ignitions would number perhaps one to two hundred. For comparison, a rule of thumb for ignitions is presented in Table 13.1, where it can be seen that for a population shaken at Modified Mercalli Intensity (MMI) VIII, there will be approximately one fire following earthquake requiring fire department assistance for each 10.5 million sq. ft. of floor area—that is, for approximately 7,000 single family dwellings, or a residential population of about 25,000.

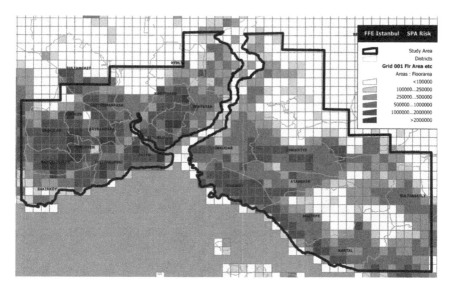

Fig. 13.2 Total Building Floor Area (sq. m) per 0.01 degree grid cell, Istanbul region

These are ignitions that require fire department response—there will be other, usually minor, ignitions that are suppressed immediately by citizens and typically not even reported.

The cause of these ignitions would likely be similar to causes in the 1994 Northridge earthquake, which is the best US data set for recent fires following an earthquake—about half of all ignitions would be electrical related, a quarter gas-related, and the other due to a variety of causes, including chemical reaction. Also based on the Northridge experience, about half of all ignitions would typically occur in single family residential dwellings, with another 26% in multi-family residential occupancies—that is, about 70% of all ignitions occur in residential occupancies. Educational facilities would be a small percentage of all ignitions (3% in Northridge), and most of these are due to exothermic reactions of spilled chemicals in chemistry laboratories.

A particular concern is oil refineries, tank farms and related energy facilities. When strongly shaken, oil refineries and tank farms have typically had large fires which have burned for days. Examples include the Showa refinery in the 1964 Niigata (Japan) earthquake, the Tüpraş refinery in the 1999 Marmara (Turkey) Earthquake (Fig. 13.4) and the Idemitsukosan Hokkaido refinery fire in the 2003 Tokachi-oki earthquake.

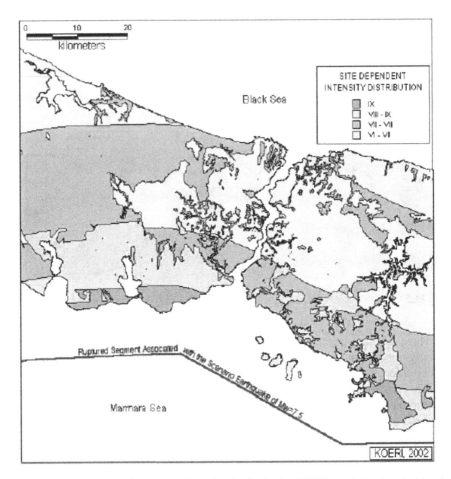

Fig. 13.3 Site dependent deterministic intensity distribution for M7.5 Scenario Earthquake (Ansal 2003)

Table 13.1 Approximate Ignition Rate versus MMI (Scawthorn et al. 2005)

MMI	VII	VIII	IX	X
1 Ign. Per million sq. ft. of Building Floor Area	18	10.5	4.5	1.5

13.2.2 Communications/Water Supply

The performance of lifelines, such as water supply, gas, electric power, communications and transportation, is integral to the fire following earthquake process.

Water supply may be severely impacted, depending on the scenario event. Generally, only the water distribution system is relevant to the fire following earthquake process. Water pressure will drop in some portions of the more heavily shaken area

Fig. 13.4 Tüpraş petroleum refinery, fire following the 1999 Marmara (Turkey) earthquake (Scawthorn 2000). Photo by G. Johnson

due to pipe breaks and tank failures. Water distribution system failure is a function of pipe material, diameter, and soil type, which for Istanbul is shown in Fig. 13.5.

Gas-related ignitions typically account for about 25% of the total number of ignitions. The Istanbul Gas Company (IGDAŞ) system is a relatively new system, Fig. 13.6, and is considered likely to suffer relatively less damage compared with older systems, such as in California (even though these systems are devoting considerable resources to modernization). IGDAŞ also has some capacity for remotely shutting trunk line valves. Nevertheless, the overall impact is that gas related ignitions may be somewhat lower than in California or Japan, due to the relatively light damage the IGDAŞ system is likely to sustain.

Communications systems, particularly telephone, will sustain some damage but not enough to reduce functionality following the scenario event. However, saturation will reduce functionality to a great degree, for several hours or more. This lack of telephone service will result in delayed reporting, with consequences as discussed above.

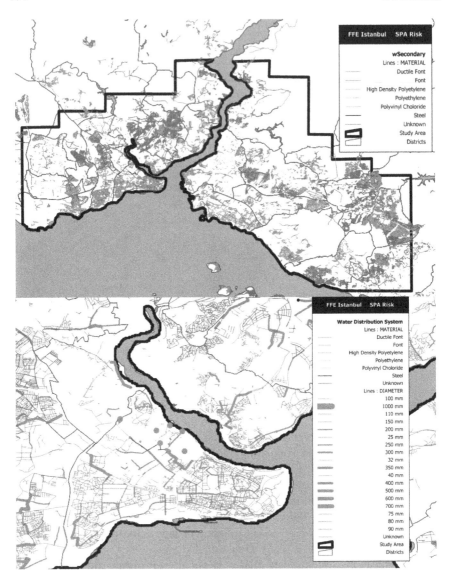

Fig. 13.5 (top) Istanbul water distribution system, (bott) detail, showing pipe materials and diameters

The transportation system most relevant to fire following earthquake is the road network, which are most vulnerable at bridge crossings. However, for Istanbul the distribution of fire stations is such that road blockage is unlikely to be a major factor, Fig. 13.7.

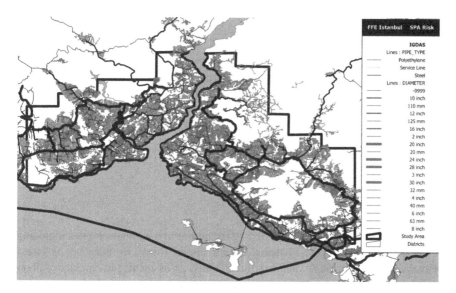

Fig. 13.6 IGDAŞ gas distribution system

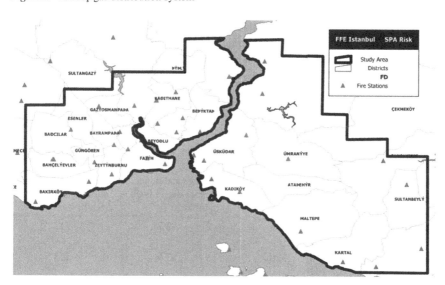

Fig. 13.7 Istanbul fire station locations

13.2.3 Fire Spread

Depending on the specific event, only a very few of Istanbul's initial few hundred ignitions will develop into conflagrations. The precise number varies with wind speed and other factors and is difficult to summarize, but an idea can be gained by again assuming a uniform PGA distribution using San Francisco, California, as a study area, Fig. 13.8. The number can be seen to be significantly less than the total number of ignitions.

13.3 FFE Risk for Several Cities

The above methods have been applied by the author to a number of cities including San Francisco, Los Angeles, Istanbul, Montreal and Tokyo, for purposes as varied as insurance underwriting, water supply reliability assessment, and emergency planning. In these studies, methods vary from fully deterministic, in which earthquake, wind and other factors are fixed and an average number of fires and associated loss are estimated, to fully probabilistic, in which uncertainty in all relevant factors is considered, and a full probabilistic distribution of losses is analyzed. Other studies are intermediate between these two extremes, varying only several key factors. Monte Carlo and other simulation techniques are commonly employed in these analyses. An example of a relatively deterministic analysis is shown in Fig. 13.9 and Table 13.2, for the city of San Francisco, California (ATC-52-1 2010).

Table 13.2 Average damage caused by fire following the scenario earthquakes, San Francisco (ATC-52-1 2010)

Scenario	Shaking damage ($ billions)[a]	Average additional damage due to fire[b] ($ billions)	Shaking plus fire damage[c] ($ billions)
Hayward Fault, Magnitude 6.9	$14	$2.7	$17
San Andreas Fault, Magnitude 6.5	$20	$3.0	$23
San Andreas Fault, Magnitude 7.2	$30	$4.3	$34
San Andreas Fault, Magnitude 7.9	$48	$5.8	$54

Notes
[a]These figures include direct damage to buildings from shaking and ground failure, in 2009 dollars
[b]These figures are averages for the many analyses with varying circumstances and do not double count shaking damage (i.e., burning rubble). Results are in 2009 dollars
[c]In 2009 dollars. Numbers in table have been rounded, which can make totals differ from sum of columns or rows

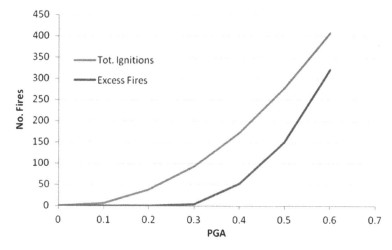

Fig. 13.8 Average no. ignitions and fires for Istanbul assuming uniform peak ground acceleration (PGA)

Another example is shown Fig. 13.10 for Montreal Canada, for three earthquake scenarios varying from magnitude 6.5–7. Accounting for fire department response, water system damage, weather and other conditions, the growth and ultimate final burnt area of fires were estimated and in summary found to result in losses of between $10 billion and $30 billion, Table 13.3 (Scawthorn 2019). These are median estimates—there are smaller probabilities of greater or less damage and the range is a function of the specific earthquake scenario (i.e., location and magnitude), time of day, weather and other factors, as shown in Fig. 13.11.

13.4 FFE Mitigation

Mitigation of fire following earthquake has been extensively discussed elsewhere (Scawthorn et al. 2005), so that only some limited recommendations are provided here, structured according to opportunities for improving fire department response and water service reliability and reducing building post-earthquake fire vulnerability.

Fig. 13.9 San Francisco showing one realization of ignitions for a M7.9 earthquake on the San Andreas fault

Table 13.3 Median results (in billions C$), fire following earthquake, Montreal Canada (Scawthorn 2019)

	S1 MC1 Mw6.5	S2 NW2 Mw7.0	S3 SW3 Mw7.0	
Deterministic	$4,215	$14,706	$12,558	Single realization, no uncertainty, daytime mild weather (20 °C, 5 km/h wind, 70% relative humidity)
Comprehensive	$11,723	$29,646	$30,655	100 realizations of spatially correlated ground motions, daytime mild weather (i.e., no uncertainty on weather)
Stochastic	$11,766	$27,653	$29,453	500 realizations of spatially correlated ground motions, uncertainty on weather and time of day

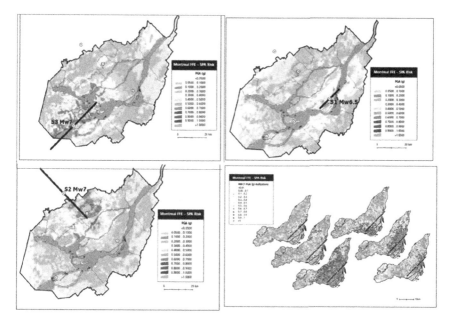

Fig. 13.10 Montreal Canada, three earthquake scenarios (upper right is Mw 6.5 in CBD, left figures are M7 to NW or SW, fault rupture shown as black line). Lower right is multiple realizations of M6.5 event considering ground motion correlation (Scawthorn 2018)

13.4.1 Fire Station Vulnerability

Fire stations and other facilities in seismically hazardous areas require immediate post-earthquake functionality and should be evaluated according to modern methods for essential facilities. In many cities, fire stations can be as much as 100 years old, and in many cases were built before 1980, which is generally considered the beginning of modern seismic design. Fire station seismic vulnerability and its effects, directly on firefighter health and safety and indirectly on the ability of the fire apparatus to respond, has long been recognized and is now well quantified, Fig. 13.12. San Francisco, Los Angeles, Vancouver, B.C., Seattle and other cities have spent millions of dollars reinforcing and replacing aged stations, specifically due to seismic vulnerability concerns (Figs. 13.12 and 13.13).

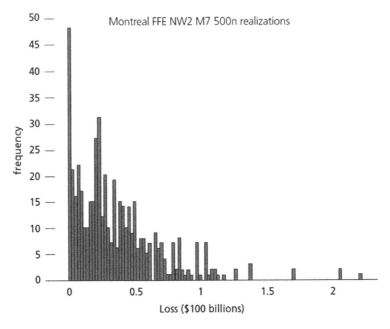

Fig. 13.11 Scenario 2 (Mw7.5) histogram of 500 realizations: mean $36 billion and median $27.7 billion (Scawthorn 2019)

13.4.2 Firefighting Water Capacity

Most cities lie on or near a shore, bay, river or other body of water, as San Francisco did in 1906—indeed, San Francisco had direct access to the largest body of water on earth—yet burned for three days due to lack of firefighting water. Why? Because, similar to the Ancient Mariner, while 'water water [was] everywhere', San Francisco could not move it to where the fires were. A similar situation exists in many cities today, which would have significant difficulty in pumping/relaying water. This situation is not unique but is also not acceptable.

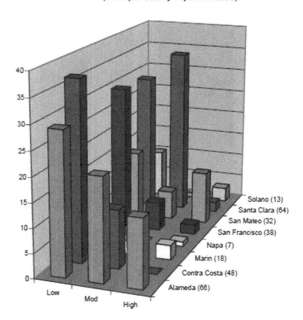

Figure 6
Number of Stations per County that are Low, Moderate and High Risk
(Total per county in parentheses)
40
35
30
25
20
15
10
5
0
Low
Mod
High
Solano (13)
Santa Clara (64)
San Mateo (32)
San Francisco (38)
Napa (7)
Marin (18)
Contra Costa (48)
Alameda (66)

Fig. 13.12 San Francisco Bay Area fire station vulnerability (Bello and Bott 2006)

Alternative water supply sources need to be better identified, and access and water transport capabilities enhanced. Large diameter hose (LDH) systems, comparable to San Francisco Fire Department's or Vallejo FD's Portable Water Supply System (PWSS) Fig. 13.14 or Vancouver B.C.'s new LDH hose reel system, Fig. 13.15, should be developed on a regional basis. Note that a PWSS has wider applicability than just earthquake—it can be used in the case of water main breaks to provide potable supply, for wildfires and for dewatering of flooded areas. See Scawthorn (2011) for further details.

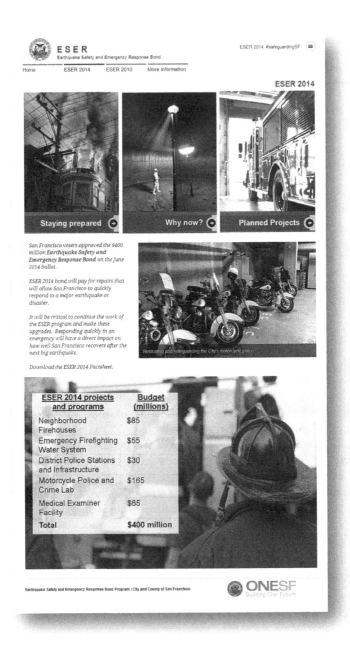

Fig. 13.13 San Francisco 2014 $400 million Earthquake Safety and Emergency Response Bond

Fig. 13.14 Example of LDH system (Vallejo FD): unit on left is a hose tender with monitor, carrying 5,000 ft. (1,538 m) of 5″ (125 mm) hose; unit on right is Hydrosub, a hydraulically driven detachable pump. The pump head can pump 1,500 gpm (6,000 lpm) up to 20 m vertically from a bridge or other point. Here it shown pumping from San Francisco Bay. See Scawthorn (2018) for more details

13.5   Concluding Remarks

In selected urban regions of high seismicity and wood building inventory, fire following earthquake is a significant but little recognized risk. Analyses in many regions has quantified this risk and led to significant mitigation which has reduced the risk. Once accomplished however, vigilance is required to maintain the mitigation measures.

Fig. 13.15 Vancouver (B.C.) Fire and Rescue new LDH reel system—each trailer-mounted motorized reel carries 6000 ft. of 6″ hose. *Photo* Scawthorn (2019)

Acknowledgements The Turkish Catastrophe Insurance Pool (TCIP) is thanked for the sponsorship and the Organizing Committee for their efforts to hold the International Workshop on Advances in Assessment and Modelling of Earthquake Loss in Istanbul in November 2019. Prof. Mustafa Erdik is particularly thanked for his leadership in this and many other earthquake risk reduction efforts. Willis Re and in particular Mr. Iman Karimi, are thanked for their sponsoring of studies of fire following earthquake for Istanbul.

References

Ansal AE (2003) Earthquake master plan for Istanbul. pp. 569. Boğaziçi University, Istanbul Technical University, Middle East Technical University, Yildiz Technical University, for the Metropolitan Municipality of Istanbul, Planning and Construction Directorate, Geotechnical and Earthquake Investigation Department, Istanbul

ATC-52–1 (2010) Here today—here tomorrow: the road to earthquake resilience in San Francisco, potential earthquake impacts. In: San Francisco: Prepared for the Department of Building Inspection, City and County of San Francisco, under the Community Action Plan for Seismic Safety (CAPSS) Project by the Applied Technology Council

Bello M, Bott J (2006) San Francisco bay area fire stations—Seismic risk assessment. In: Proceedings of the 8th U.S. national conference on earthquake engineering, Paper No. 001662, San Francisco

History of Istanbul. https://www.istanbul-istanbul.net/history/history10.htm

Scawthorn C (2000) The Marmara, Turkey Earthquake of August 17, 1999: Reconnaissance Report, MCEER Tech. Rpt. MCEER-00–0001, Multidisciplinary Center for Extreme Events Report, State University of New York at Buffalo

Scawthorn C (2011) Water supply in regards to fire following earthquakes. In: 173. Berkeley: Pacific Earthquake Engineering Research Center, College of Engineering, University of California, sponsored by the California Seismic Safety Commission. www.seismic.ca.gov/pub/CSSC_2011-02_WaterSupply_PEER.pdf with four page summary at https://peer.berkeley.edu/publications/peer_reports/reports_2011/Fire%20Following%20Earthquake-online-view-layout-sm.pdf

Scawthorn C (2018) Fire following the Mw 7.05 Haywired earthquake scenario. In: Detweiler ST, Wein AM (eds.) The HayWired Earthquake scenario—engineering implications (U.S. geological survey scientific investigations report 2017–5013–I–Q: Washington)

Scawthorn C (2019) Fire following earthquake in the Montreal Region. In: 100 pp. Toronto. https://www.iclr.org/wp-content/uploads/2019/11/Montreal-fire-following-earthquake_E.pdf : Prepared for Institute for Catastrophic Loss Reduction

Scawthorn C, Khater M (1992) Fire following earthquake—conflagration potential in the greater Los Angeles San Francisco Seattle and Memphis Areas. prepared for the Natural Disaster Coalition by EQE International, San Francisco, CA. [EQE's study for the Natural Disaster Coalition, Fire Following Earthquake in the Greater Los Angeles, San Francisco, Seattle and Memphis Areas, is available from the National Committee on Property Insurance, 75 Tremont Street, Suite 510, Boston, MA 02108–3910, (617. 722–0200.]

Scawthorn C, Yamada Y, Iemura H (1981) A model for urban postearthquake fire hazard. Disasters 5(2):125–132

Scawthorn C, Eidinger JM, Schiff AJ (2005) Fire following earthquake. Technical council on lifeline earthquake engineering monograph No. 26, American Society of Civil Engineers, Reston, 345pp

Index

© The Editor(s) (if applicable) and The Author(s) 2021
S. Akkar et al. (eds.), *Advances in Assessment and Modeling of Earthquake Loss*,
Springer Tracts in Civil Engineering,
https://doi.org/10.1007/978-3-030-68813-4

Printed in the United States
by Baker & Taylor Publisher Services